Ecomobilities

Environment and Society

Series Editor: Douglas Vakoch

As scholars examine the environmental challenges facing humanity, they increasingly recognize that solutions require a focus on the human causes and consequences of these threats, and not merely a focus on the scientific and technical issues. To meet this need, the Environment and Society series explores a broad range of topics in environmental studies from the perspectives of the social sciences and humanities. Books in this series help the reader understand contemporary environmental concerns, while offering concrete steps to address these problems.

Books in this series include both monographs and edited volumes that are grounded in the realities of ecological issues identified by the natural sciences. Our authors and contributors come from disciplines including but not limited to anthropology, architecture, area studies, communication studies, economics, ethics, gender studies, geography, history, law, pedagogy, philosophy, political science, psychology, religious studies, sociology, and theology. To foster a constructive dialogue between these researchers and environmental scientists, the Environment and Society series publishes work that is relevant to those engaged in environmental studies, while also being of interest to scholars from the author's primary discipline.

Recent Titles in the series

Ecomobilities: Driving the Anthropocene in Popular Cinema, by Michael W. Pesses
Embodied Memories, Embedded Healing: New Ecological Perspectives from East Asia, edited by Xinmin Liu and Peter I-min Huang
Wetlands and Western Cultures: Denigration to Conservation, by Rod Giblett
Sustainable Engineering for Life Tomorrow, edited by Jacqueline A. Stagner and David S. K. Ting
Nuclear Weapons and the Environment: An Ecological Case for Nonproliferation, by John Perry
Portland's Good Life: Sustainability and Hope in an American City, by R. Bruce Stephenson
Conservation, Sustainability, and Environmental Justice in India, edited by Alok Gupta
Environmental and Animal Abuse Denial: Averting Our Gaze, edited by Tomaž Grušovnik, Reingard Spannring, and Karen Lykke Syse
Living Deep Ecology: A Bioregional Journey by Bill Devall, edited with an introduction by Sing C. Chew
Environmental and Animal Abuse Denial: Averting Our Gaze edited by Tomaž Grušovnik, Reingard Spannring, and Karen Lykke Syse
Secular Discourse on Sin in the Anthropocene: What's Wrong With the World? by Ernst M. Conradie

Ecomobilities

Driving the Anthropocene in Popular Cinema

Michael. W. Pesses

LEXINGTON BOOKS
Lanham • Boulder • New York • London

Published by Lexington Books
An imprint of The Rowman & Littlefield Publishing Group, Inc.
4501 Forbes Boulevard, Suite 200, Lanham, Maryland 20706
www.rowman.com

6 Tinworth Street, London SE11 5AL, United Kingdom

British Library Cataloguing in Publication Information Available

Library of Congress Cataloging-in-Publication Data Available

ISBN 978-1-4985-9819-4 (cloth)
ISBN 978-1-4985-9820-0 (electronic)

Contents

Acknowledgments

I wish to give a sincere thank you to all of those, anonymous or otherwise, who have given constructive criticism of this work as it has evolved since its inception. Notably, Eve Oishi, Darrel Moore, and Nadine Chan provided guidance and support that was most appreciated.

As this wound up mainly being written during the COVID-19 pandemic, I must give the majority of my thanks to Sarah, Jack, Zoe, Alex, and Max who were stuck with me while I tried to maintain a normal work schedule despite nothing being normal about the past few years. I promise the next project will be less bizarre.

Introduction

Driving the Anthropocene

IT'S NOT LOOKING GOOD . . .

In May of 2021, the World Meteorological Organization (WMO) wrote a press release announcing that the odds were increasing that average annual temperatures for the globe will reach 1.5°C above pre-industrial levels in the next five years.[1] That number is a threshold that most countries of the world have been working to stay below as it would keep the damage of a warming planet as minimal as possible at this point. Sea levels will continue to rise due to melting glaciers, for example, but staying under 1.5°C means that they would rise at least ten centimeters less than if we only met our previous goal of staying between 1.5–2°C.[2] In effect, keeping our global emissions down won't reverse or even stop climate change from happening, but our lives won't be as terrible as they could be if we were to do nothing.

What makes this news all the more depressing is that despite the global impact of the COVID-19 pandemic on travel, transportation, and industry, from a climate change perspective we are no better off than we were before the lockdowns. Yes, deadly air pollutants like nitrogen dioxide (NO_2) dropped during this moment in time, improving the air in cities and saving thousands of lives, but, stressed the WMO, we are still hurtling toward crossing that 1.5°C threshold.[3]

I am not interested in simply rehashing the depressing data and predictions though. The last thing we need at this moment is more existential dread. Rather, my goal is to question how some countries, primarily the United States, have responded to the knowledge of climate change and what could be done to limit its effects. Despite the seemingly newness of climate change and global warming as a focus of study in the twenty-first century, humans have long been aware of their ability to alter the weather through the consequences of industry.[4] Even the concept of the greenhouse effect, the mechanism through which a layer of gasses in the atmosphere helps regulate temperatures on Earth, was worked out in the early nineteenth century.[5] The first hypothesizing on global warming took place in 1896 by Swedish scientist Svante Arrhenius, though he dramatically underestimated how long it would take for humans to significantly increase the amount of carbon dioxide (CO_2) in the atmosphere.[6]

So if we have known about the possibilities of environmental degradation, global warming, and climate change for well over a century now, why have we humans been so reluctant to do anything about it? *Ecomobilities* is an effort to get at the ideology at work, especially in the twentieth and twenty-first centuries that has been at odds with the climate crisis. Specifically, I want to explore the ideological connections between automobiles, the environment, and the end of the world. I am not so much interested in the harmful emissions of the cars themselves, but instead in what I see as a modern inability to envision traveling through ecosystems, including urban ones, without the aid of an automobile. What is it about the car that makes it inseparable from modern life despite the evidence that it is so damaging for environmental health? The answer, I will argue, is inherently ideological, and something we must begin to grasp if we are to actually do anything about our future environmental health. The automobile is not the only contributor to climate change, and clearly a part of a larger capitalist system, but it is a commodity that reveals much about how we got into this crisis. To get at this ideology, I have chosen to work in popular film. This is not a work of film theory but rather an effort to use film to reflect on ideologies affecting our response to climate change. Specifically, I contend that an ideology of American automobility has influenced how we believe we ought to respond to warming temperatures and shifting ecosystems. The stories we tell reveal the lens through which we see the world.

WHAT'S IN A NAME? TROUBLING THE ANTHROPOCENE

I want to use what follows to discuss the automobile within the context of the Anthropocene, a proposed new geological epoch that marks the new atmospheric composition since the Industrial Revolution. The term itself was coined in the year 2000 by scientists Paul J. Crutzen and Eugene F. Stoermer in response to the obvious human effects on the natural systems of the Earth:

> Considering these and many other major and still growing impacts of human activities on earth and atmosphere, and at all, including global, scales, it seems

> to us more than appropriate to emphasize the central role of [hu]mankind in geology and ecology by proposing to use the term "anthropocene" for the current geological epoch. The impacts of current human activities will continue over long periods. According to a study by Berger and Loutre, because of the anthropogenic emissions of CO_2, climate may depart significantly from natural behavior over the next 50,000 years.[7]

Officially, Crutzen, Stoermer and other scholars would stop the Holocene, the epoch in which we currently exist, at the end of the nineteenth century to match it with the beginning of a global industrial revolution and thus significant fossil fuel consumption.[8] Such a break reflects the majority of climate science, including the pre-industrial temperature levels used by the Intergovernmental Panel on Climate Change (IPCC) and the WMO, which refers to a specific reference period from 1850 to 1900, ideally representing a time before we humans began pumping greenhouse gases into the atmosphere in previously unseen numbers.[9]

While the term "Anthropocene" has been accepted and widely used in social and cultural studies of Earth's changing climate, it is not the only way to describe a new epoch on this planet. The most useful alternative is Andreas Malm's term "Capitalocene," which, rather than replace Anthropocene, serves to remind us as to just how exactly human action has resulted in planetary change. For Malm, the use of *anthropos* is an "indefensible abstraction" when it comes to making sense of our changing climate as it places all of humanity as both a monolithic cause and victim of the impending ecological crisis.[10] "Unlikely to gather anything like a consensus behind it, a more scientifically accurate designation, then, would be 'the Capitalocene.' This is the geology not of [hu]mankind, but of capital accumulation."[11] Since both resource extraction and quest for profit have driven the increase in greenhouse gasses, it stands to reason that the capitalists are to blame for this mess. To make matters worse, economic inequality means that not all humans will suffer through this crisis equally: "there *will* be lifeboats for the rich and privileged, and there will *not* be any shared sense of catastrophe."[12] Why blame humanity for this new epoch when an elite few brought us into it?

The Capitalocene has been taken up by Jason W. Moore, who has extensively explored the interweaving of nature and capital rather than a more generalized study of nature and humans. Moore argues that the dualism of Nature/Society has allowed a violence that "drips with blood and dirt" in that nature has been rendered passive.[13] For Moore, the urgent move is to think of how the dualism of nature and society has limited our abilities to understand the climate crisis. By conceiving of capitalism-in-nature, we can begin to think of how capitalism works "*through*, rather than upon *nature*."[14] Moore pushes for an ontological break in the separation of nature and society by studying both within a historical materialist relationship:

> There has been too little investigation into how bundles of human and extra-human relations constitute modernity's historical natures, and how patterns of power and capital are producers *and* products of those natures. The conventional wisdom says that modernity makes environmental history. But is not a more relational proposition more tenable: modernity *as* environmental history?[15]

The theme running throughout Moore's work is to fight against any dualism when it comes to human society and the natural world. Even this push for a recognition of capital's role in the climate crisis is not to separate or pit capital versus nature, but rather to set them together. I see Moore's work as being evocative of Neil Smith's *Uneven Development: Nature, Capital, and the Production of Space*, in which he argues that the externality of nature from human society is connected to the rise of capitalism yet is not simply a result of capital. Smith is arguing for a dialectic, rather than mere interactions between two spheres, much as I see Moore calling for with the Capitalocene. For Smith, the relationship exists within production:

> Elements of the first nature, previously unaltered by human activity, are subjected to the labor process and re-emerge to be social matter of the second nature. There, though their form has been altered by human activity, they do not cease to be natural in the sense that they are somehow now immune from non-human forces and processes—gravity, physical pressure, chemical transformation, biological interaction. But they also become subject to a new set of forces and processes that are social in origin. Thus the relation with nature develops along with the development of the social relations, and insofar as the latter are contradictory, so too is the relation with nature.[16]

This second nature of which Smith speaks is when nature has been produced through human labor. It is not less natural in the sense that physical laws no longer apply, but it is distinctly touched by societal forces. For Smith, those societal forces infect this second nature with the political and ideological. While this dialectic is not unique to modern society, Smith contends that our relationship with nature has been inherently changed by capitalism. As Marx puts it

> nature does not produce on the one hand owners of money or commodities, and on the other hand men possessing nothing but their own labor-power. This relation has no basis in natural history, nor does it have a social basis common to all periods of human history.[17]

And it is Marx's conception of the metabolic interaction between humans and nature that Moore builds his critique of the Anthropocene. Marx's metabolic interaction is a universal human experience, "the everlasting nature-imposed condition of human existence, and is therefore independent of every form of that existence, or rather it is common to all forms of society in which human beings live."[18] All humans must live within nature to survive, but the capitalist dramatically alters the interaction. Moore sees the placing of the Anthropocene problematic as it masks capital's relationship to this interaction. Starting the epoch with the Industrial Revolution overlooks the centuries of appropriation and exploitation of nature that facilitated the shift. "The rise of capitalism launched a new way of organizing nature, mobilizing for the first time a metric of wealth premised on labor productivity rather than land productivity."[19] The rise was not during the nineteenth century as machines replaced human hands but instead during colonial exploration starting in the fifteenth century.

> Ask any historian and she will tell you: how one periodizes history powerfully shapes the interpretation of events, and one's choice of strategic relations. Start the clock in 1784, with James Watt's rotary steam engine . . . and we have a very different view of history—and a very different view of modernity—than we do if we begin with the English and Dutch agricultural revolutions, with Columbus and the conquest of the Americas, with the first signs of an epochal transition in landscape transformation after 1450.[20]

It is at this moment that Moore sees a distinct shift in the metabolic interaction that demands the dualistic relationship of nature and society. The Industrial Revolution can be read as a means to fix the organizational structure of capitalism to deal with earlier crises.[21]

I find incredible use in Malm's, Moore's, and Smith's historical materialist readings of the interlocking reality of nature, capital, and environmental degradation, especially the theorizing done by the latter two men. And yet, I want to signal a departure from this work by using Anthropos, rather than Capital, to describe the 'cene. Primarily, I see value in the Anthropocene in that it is a catchier title. Even Malm admitted above that it would be hard to make Capitalocene catch on. There is something to be said for grabbing a hold of an already popular name and using it in a more rigorous manner.

Moore and the others have made valid points about blaming the accumulation of capital rather than all of humanity for our climate crisis. And yet, I fear that the term "Capitalocene" masks just how pervasive and all-encompassing the climate crisis is. I, identifying as a radical, a leftist, surely cannot be blamed for increasing temperatures! I am fighting the capitalist bastards, after all. As Joel Wainwright and Geoff Mann write in the preface to their

Climate Leviathan, "The vast proportion of historical greenhouse gases have been emitted as byproducts of the choice and activities, not of the masses of ordinary people, but rather a wealthy minority of the world's people."[22] While their argument for a political response to climate change goes beyond simply blaming a select few, I cannot help but get the sense that middle-class North Americans and Europeans, including many academics writing about this subject are supposed to slip into the category of those without blame. I am arguing that we need to be a little more all-encompassing when it comes to the superstructural components of climate change. Following Latour, I have no wish to attack "the worker forced to travel long distances by car because she hasn't been able to find affordable housing near the factory where she works: who would dare shame her on account of her carbon footprint?"[23] But I do think we need to start dismantling such a system and that work cannot be simply to complain about elites. We must lay bare the ideologies that keep us not just reproducing the system, but unable to imagine another way forward. To stick with the name Anthropocene is not to blame humans *qua* humans for climate change but to acknowledge the implications of humans as a force of change on the planet. Not all humans must have power nor capital in order to be connected to ecological change. We can still work against class inequalities as we study societal practices across class lines.

My interest in the Anthropocene lies not in revealing the direct exploitation of nature by capital, but rather the invisible superstructural ways in which power flows to maintain the appearance that it is perfectly natural to separate ourselves from nature. The abundance of books and articles arguing for the unification of society and nature suggests that we all know very well that humans and nature are connected. It seems unnecessary to write yet another argument against separating society from nature. Instead, we need to get at why these repeated treatises on the hybridity of society/nature seem so necessary despite our best efforts.

The Capitalocene studies have done the heavy lifting, started by Marx, that theorizes the production link between human and nature. Building off these works, I want to mess around with ideology, a complicated, if not dirty, word in Marxist theory. Perhaps the fact that my interest is less in production and labor and more in ideology and media helps to explain why I gravitate to the Anthropocene rather than the Capitalocene. I am not trying to get at the exact moment in which the climate change began, but rather what keeps it going and makes it seemingly impossible to address. Many of the Anthropocene studies are looking to what this new reality means for the individual, even if addressing collective responses. The cause might be the accumulation and movement of capital, but many of the effects are felt at the individual level. In some cases that individual is an informed consumer, in others it is simply an alienated individual scared into paralysis by what the future holds.

As McKenzie Wark has pointed out in her own study of the Anthropocene, our response to climate change typically works in one of two ways. The first response invokes Mark Fisher's *Capitalist Realism*, in that we cannot imagine our way out of capitalism and thus cannot imagine fixing climate change without that system.[24] "This insists that there is no alternative, and we just have to stick with the program. If it takes the planet down with it then so be it."[25] The other option is to return to a state of society that (seemingly) existed before capitalism took over.

> The alternative narrative imagines a kind of non-technical, holistic and spiritual alternative, often drawing its images from a pre-capitalist landscape. But as was already clear to Marx, this is *capitalist romance*, a story constructed within capitalism itself as one of the byproducts of its own momentum. It is a kind of capitalist realism in negative, where we all ride bamboo bicycles, but it rarely ventures beyond an ideological mirroring of capitalist realism.[26]

Wark goes on to invoke critical theory to envision new framings of the Anthropocene to address the deficiencies of these two approaches, but I want to stick with them for a bit longer. Why is an alternative to capitalism so daunting to envision? Why must the only way to limit Earth's changing climate be to turn back the clock to a pre-capitalist state?

I am arguing that we do not fully understand the ideological work being done to make not just capitalism, but the mobilities of capitalism seem so inherently natural. The answer lies in the ideological work being done in mass culture that, as any good ideology is, is invisible to the consumer. My interest here is to specifically examine how the automobile has been represented in film as a window to an overarching ideology of American automobility. I don't just want to point out the presence of cars in films (even though I love such essays and books), but I want to trouble the very idea of the car as an ideology that affects everyday life. Until we lay bare the mobilities of our daily lives and connect them to the greater capitalist lens through which we see the world, we will never be able to imagine a new future on this planet.

I have found that it makes more sense to think in multiplicities rather than dialectics when it comes to reactions to climate change. I want to spend time with the *anthropos* to see our rather human reactions to what capital has wrought. I also find value in the various notions of hybridity and assemblage that have been offered to explain both the Anthropocene and the automobile.

Clearly, there are problems in the theorization. For example, Purdy's *After Nature*, with its leading title and constant references to a "(post-)natural world" thanks to human activity fails to take Marx's work into account, but his insistence on nature being completely inseparable from humanity is interesting.[27] That is, we have seen the arguments that one cannot separate society and nature, argued in a variety of ways from materialism to post-humanism.[28] But what Purdy does, and I think more so than many other scholars of the Anthropocene, is to insist that not only are they inseparable, but that humanity is always already in nature.

> The natural and the artificial have merged at every scale. Climate change makes the global atmosphere, its chemistry and weather systems, into Frankenstein's monster—part natural, part made. The same is true of the seas, as carbon absorption turns the oceans acidic and threatens everything that lives in them. . . . Even wilderness, that emblem of untouched nature, persists where lawmaking and management create it, artificial testament to the value of natural things.[29]

Clearly, much of this is not new ground, but the continued insistence of an inability to go to a humanless spot on the planet is emblematic of the Anthropocene. As I am interested in the automobile's role in all of this, the link between hybrid driver-car and the hybrid capital-ecosystem provides a previously untapped space of research. Both pairings also offer a chance to trouble the ideological work that renders them invisible in everyday life. What I want to challenge though, if we really are going to push forward with the Anthropocene, is this notion of "the artificial" to describe things that have come into being in the last few centuries. If we are truly in the Anthropocene then we are in an epoch of assemblage not rigid classifications.

TROUBLING IDEOLOGY

Ideology as it will be used here is derived from Marx, who, as summarized by Althusser, saw ideology as "the system of ideas and representations which dominate the mind of a [hu]man or a social group."[30] While such a summation might seem benign, ideology in the Marxist tradition has had a negative connotation, in that it represents a class-based view of reality. As Smith puts it, "Ideology is not simply a set of wrong ideas but a set of ideas rooted in practical experience, albeit the practical experience of a given social class which sees reality from its own perspective, and therefore only in part."[31] The ideas that constitute ideology are, if not verifiably false, incomplete perceptions of the world.

The key aspect of a Marxist, and thus materialist conception of ideology, is that these ideas are not something plucked from the *aether*. In the first part of "The German Ideology," Marx explicitly contends that ideas are formed through material processes:

> We do not set out from what [people] say, imagine, conceive, nor from [people] as narrated, thought of, imagined, conceived, in order to arrive at [people] in the flesh. We set out from real, active [people], and on the basis of their real life-process we demonstrate the development of the ideological reflexes and echoes of this life-process. The phantoms formed in the human brain are also, necessarily, sublimates of their material life-process, which is empirically verifiable and bound to material premises. Morality, religion, metaphysics, all the rest of ideology and their corresponding forms of consciousness, thus no longer retain the semblance of independence. They have no history, no development; but men, developing their material production and their material intercourse, alter, along with this their real existence, their thinking and the products of their thinking. Life is not determined by consciousness, but consciousness by life.[32]

Marx's radical move here is to critique the acceptance of ideas as being superorganic objects to be plucked from the sky to inform human society. Ideas come from the material actions of humans interacting with one another. To understand ideas one must look at the context in which they form. Further, Marx later claims that "the ideas of the ruling class are in every epoch the ruling ideas: i.e., the class which is the ruling *material* force of society, is at the same time its ruling *intellectual* force."[33] Not only are ideas connected to material phenomena, a dominant group forces those ideas onto the underclasses.

Marx's ruling class/ruling ideology argument was taken up in the late twentieth century, first by Althusser in which he uses ideology to explain how labor and power relations are reproduced in a capitalist economy. Marx's ruling class/ruling ideas thesis is rejected for its repressive nature. For Althusser, ideology is not the real relationship between members of a society, but the imaginary relation of individuals to the real relations in which they exist.[34] So ideology is an idea, but one that springs forth from the material world. Where Marx contends that ideology has no history, Althusser explains that ideology is better thought as having "no history *of its own*,"[35] meaning ideology is tied both materially and historically to subjects. "An ideology always exists in an apparatus, and its practice, or practices. This existence is material."[36] Within this dialectic, Althusser quite bluntly removes the notion of "ideas" and replaces it with "practices, rituals, ideological apparatus."[37] These practices and rituals cannot be totally repressive however; subjects would not accept the ideology. An "Ideological State Apparatus" (as opposed to a repressive apparatus) allows for the continued reproduction of relations of production through the productive interpellation of individuals as subjects.[38]

Althusser is directly critiqued in Stuart Hall's essay, "Signification, Representation, and Ideology: Althusser and the Post-Structuralist Debates" in which he argues that we must go further in rejecting the repressive nature of Marx's ruling class/ruling ideas hypothesis. What Hall is most interested in "is how a society *allows* the relative freedom of civil institutions to operate in the ideological field, day after day, without direction or compulsion by the State; and why the consequence of that 'free play' of civil society, though a very complex reproductive process, nevertheless consistently reconstitutes ideology as a 'structure in dominance.'"[39] Hall argues that the State is far from an all-powerful repressive force, and yet, ideological reproduction benefits and reproduces the State. While Hall does evoke and agree with Michel Foucault, he works toward a middle ground between the Marxist repressive State and Foucauldian multiplicities of dispersed power. Hall's problem lies in the fact that while there is no singular State apparatus, a State does exist. For Hall, the answer lies in Gramsci's concepts of hegemony, consent, and common sense to wrestle with the complexities of class and ideology. Where both Foucault and Althusser are quick to eschew ideas and what happens in the mind of a subject, Hall pushes for a mental space of ideology, connected to, but distinct from material relations. The common sense ideas in society, those that become naturalized to the point of being thought of as "that's just how life is," are precisely what ideology is all about. "The point at which we lose sight of the fact that [common] sense is a production of our systems of representation is the point at which we fall, not into Nature but into the naturalistic illusion: the height (or depth) of ideology."[40] That interface of the natural and the social is where ideology exists. To change ideology then, is, as Hall argues, a process of "articulation" in which ideological signs are altered and given new common sense meanings.

Perhaps what is most important, though, is that ideology is invisible in everyday life. If ideology has been made natural in the Gramscian sense, as Hall contends, then the average subject would not question social relations. The problem with such a line of inquiry though is that it tends toward a simple answer of pulling back the curtain to see reality. *The Wizard of Oz* depicts this quite literally. As the dog Toto pulls on the green curtain, revealing not the great and powerful Oz but a small, flustered man furiously pulling levers, Oz (Frank Morgan) shouts, "pay no attention to that man behind the curtain!" A more modern version of this takes place if we turn to *The Matrix*. *The Matrix* refers to the false consciousness in the form of a computer program that allows docile human bodies to be harvested for energy by their machine overlords. The humans stuck in goo filled pods have no idea about reality and instead go through life as if it were still the late 1990s. At a key moment in the film, Morpheus (Laurence Fishburne) offers Neo (Keanu Reeves) either a red pill, which will enlighten him and "pull back the curtain" or a blue pill which will maintain the ideological mask and he will return to his life as if nothing happened. Neo takes the red pill and is now free to see reality for what it is, though he soon learns how to reenter the Matrix to work toward its destruction. This "demasking" as Slavoj Žižek puts it,[41] misses just how powerful ideology can be and how often real resistance never materializes. Simply knowing that ideology is not synonymous with reality does not eradicate the ideology. If anything, pulling back the curtain can reinforce and help reproduce the ideology. "They know very well how things really are, but still they are doing it as if they did not know. The illusion is therefore double: it consists in overlooking the illusion which is structuring our real, effective relationship to reality. And this overlooked, unconscious illusion is what may be called the *ideological fantasy*."[42] So even if we can see through the illusion of ideology, this ironic distance to ideology that acknowledges the "curtain" or "program" thus allows one to still operate within the ideological framework.

The real move, then, is to work to expose the ideology at hand, but also to not slip into an ironic distance that works against making real changes for those negatively affected by that ideology. The ideology of American automobility, however, has yet to be fully exposed.

AUTOMOBILITY, AMERICAN, AND OTHERWISE

"Automobility" is a rather loaded term, especially when one adds American to the front. In what follows, I will define American automobility as the cultural, social, and technical aspects of the predominant method of autonomous movement within the distinct relations of power of the United States from the late nineteenth century to the present. In simpler terms, we are looking at the material and ideological ways in which individual Americans move. Automobility need not be restricted to the automobile, but the automobile is clearly the dominant form of personal transportation in the United States. I take the prefix American from Cotten Seiler's *Republic of Drivers: A Cultural History of Automobility in America*. While his history is focused on the United States, the term American goes beyond national borders to "signify myth, transmit ideology, and confer power."[43] Seiler's work is a sustained look at how the automobile was worked into everyday life in America and, he argues, destroyed a sense of community through the individualizing of drivers. Quoting Henri Lefebvre, Seiler describes a typical scene on an American interstate highway, "what I see in my window is the republic of drivers in a moment of plenitude: the drivers move freely in their sociality of 'simultaneity without exchange'; and the landscape through which they pass orders and enables their movements."[44] Not only do drivers exist without meaningful interaction with one another, the very act of driving in the

United States is a Foucauldian *dispositif* of control and discipline. According to Ann Stoler, Michel Foucault's concept of *dispositif* is usually translated as "social apparatus" which misses the spatiality at work. "A *dispositif* . . . is not a thing but the system of connections among this ensemble of arrangements."[45] The mobility inherent in driving in the United States makes such a network of power relations a bit easier to grasp. Also following Foucault, Seiler argues that American automobility exists within a specific moment in time, with a specific genealogy. Interestingly, he does not use the often cited Interstate System of 1956 as a break or rupture in American automobility, but rather a continuation that "merely dedicated a larger share of resources to a covenant with automobility that was many decades old."[46] This "covenant with automobility" was about the notion of being both modern and a free subject within the larger American society. Seiler situates the rise and peak of American automobility from 1895 to 1961, beginning with the advent of industrialization and the scientific management of the Taylorization of labor which led to a crisis of individualization. While Seiler is a post-structuralist in the sense of the individual being an invention during the Enlightenment, he sees the concept of a *lack* of individualization as a moment in American history that led to the development of automobility.[47] The robotic, scientific motions disciplining workers did not fully quash the individual and the concept saw a resurgence in the 1920s and again at the start of the Cold War. Automobility allowed America to recapture a sense of masculinity and individualism while still promoting industrial economic growth. Roads were built and cars were sold in terms of economic growth, national defense, and as a means to connect the country. Masculinity and individualism were simply "common sense" outcomes of an autonomously mobile population.[48] Automobility was made natural within networks of Foucauldian power in both the discourse and materiality of the automobile. Seiler claims that "automobility comprises a 'multilinear ensemble' of commodities, bodies of knowledge, laws, techniques, institutions, environments, nodes of capital, sensibilities, and modes of perception."[49] Automobility is clearly a white, middle to upper class, male apparatus, though Seiler examines other groups existing in American automobility of the first half of the twentieth century. Both women and African Americans saw automobility as a means to independence and freedom, but Seiler argues that through driving these individuals were made subjects of the dominant ideology. Women using automobiles were seen as necessary to reproducing a domestic capitalism, yet they were working within masculine spaces of the car and road.[50] African American motorists faced racial injustices on the highways, which they attempted to mitigate with guidebooks like *Travelguide (Vacation & Recreation without Humiliation)* and *The Negro Motorist Green Book.* Seiler argues that while automobility did not provide access to a nonracial subjectivity as some had hoped, the development of the Interstate System aided in escaping the Jim Crow laws of the South.[51] Further, the subcultures of automobility like low-riders should not be seen as a resistance to automobility, but rather something like Raymond Williams' "alternative" inflections of "the dominant hegemonic practice."[52]

While I agree with Seiler's rise of American automobility, placing its zenith in 1961 misses the events of the late twentieth and early twenty-first centuries that increased an American reliance on automobility despite clear environmental concerns. The OPEC energy crisis of the 1970s was a moment that should have destroyed American automobility had it peaked a decade prior. The cutoff of oil from the OPEC countries led to higher costs and rationing. Such a moment of danger could have led the American public to discuss the merits of continued individual mobility; rather we bought smaller, more efficient cars until the oil flowed and the price went back down. Fuel economy standards born of the OPEC crisis, as well as hybrid, fuel cell, and electric technologies seen as necessary for a warming climate show how we refuse to abandon the automobile, but rather try to use it to fix our environment. Elon Musk just launched a Tesla automobile into space; how could we have peaked over a half-century ago? In what follows, I will be arguing that American automobility's peak has yet to arrive.

If we work from the assumption that American automobility is in fact an example of a Foucauldian *dispositif*, and I think that is a good starting point, more care must be given to the subjugation of the American driver. In *Discipline and Punish*, Foucault studies the *dispositif* at the site of the body; subjects are made through repetitive movements and surveillance.[53] Seiler argues that driving is the practice that makes the subject, yet he spends most of his work on the construction of an ideology that justified producing the automobility *dispositif*.[54] To be fair, invoking Foucault to study ideology would appear to be doomed from the start; Foucault himself shot down the validity of ideology as a concept.[55] I think what must be done is to connect the material existence of a networked apparatus of power such as American automobility to how such power relations can be accepted within a subject's mind. If I know that automobility is bad for my health, the environment, my personal finances, and so on, why do I still submit to it? The answer must lie in ideology.

American automobility is in fact an ideology. More than simply a means for traveling from point A to point B, American automobility has become a means for interpreting the world. Again, returning to Althusser and Marx, ideology is "the system of ideas and representations which dominate the mind of a [hu]man or a social group."[56] As will be shown throughout this book, the use of an individual automobile for mobility has become the common sense means to travel for those subjects of American automobility.

The car as the best tool for the job is one component of this ideology, but it does not directly translate into an example of ecomobility. Automobility is at its most natural when it becomes the *only* way to travel through a specific ecosystem.

In Nicolas Winding Refn's 2011 film, *Drive*, cars are more than objects of desire in the film. The driver (Ryan Gosling) is a Hollywood stunt driver, a mechanic, and an incredibly competent getaway driver. He understands cars and drives them with a skill well beyond the average motorist. Even Gosling's (intentionally) robotic performance suggests a connection with the machines different from one held by the average motorist. The cars themselves are full of muscle; the sound of the revving and shifting engines evokes the thrill of machine-assisted speed. It would be easy to point to such a film as an example of ecomobility, but the cars never transcend the role of a tool for movement. The automobile is a tool for earning money, for impressing a woman and her son, for exacting revenge, but we are always aware that it is a tool. The same plot could exist while replacing cars and driving within another skill that could assist criminals, like computers and hacking. The fact that we are always aware of the automobiles prevents them from being made natural, and thus from reinforcing ideology. This is not a failure of the film; *Drive* simply isn't that kind of film. Despite the exciting car chases and getaways, *Drive* does not reproduce the ideology of American automobility. We may enjoy the loud engines and fantasize about driving with such competence, but we are not subjugated through such a film.

The original *The Karate Kid* is a wonderful example of how film contributes to the production of American automobility's subjects, as well as how certain mobilities are necessary for certain environments. The film, on the surface, tells the story of young Daniel LaRusso (Ralph Macchio) who learns Karate from the wise Mr. Miyagi (Pat Morita) in order to fit into his new home in Southern California, stop the bullies, and get the girl. The martial arts, while entertaining, are not the most crucial aspect of *The Karate Kid*, however. This is a film about a naturalized automobility in a distinctly American sense.

The film opens with Daniel LaRusso (Ralph Macchio) moving across country with his mother in their old green station wagon. Again, the car is a tool; the LaRussos could have flown or taken a train to their new Southern California home. Daniel does not yet have his driver's license, which means he either must rely on his mother to drive him from place to place or use his bicycle. The ideology exists in Daniel's bicycle, a transport mode that is deviant in the ecosystem that is the San Fernando Valley in the 1980s. I myself, as a boy living in California and watching the film when it was first released, thought that Daniel's bicycle was actually pretty cool but even my young mind knew that there were other modes of transport to which one needed to aspire.[57] The Cobra Kai bullies, who are also karate experts, ride motorcycles that are not explicitly stated as better than Daniel's bicycle, but are shown to be superior throughout the film. In one scene, Daniel rides his bicycle home at night when the headlights and buzzing of the motorcycles approach. Johnny, the leader of the bullies (William Zabka) is mocking Daniel for wanting to learn karate and knocks him off of a hillside so he can learn his first lesson, "how to take a fall." Daniel is humiliated, primarily because he lacks the access to an accepted mobility.

Yes, Daniel will train with Mr. Miyagi and ultimately defeat the Cobra Kai bullies in a tournament. Daniel's life really turns around, however, on his sixteenth birthday in which Miyagi gives him one of the many restored old cars he owns. Now that Daniel has a car, and a nice one at that, he gets his girlfriend back and the whole tone of the film shifts to indicate that despite his struggles, he can win the All Valley Under 18 Karate Tournament. Such a film lets slip the Gramscian common sense of American automobility which is all but invisible in a film like *The Karate Kid*. The film viewer is not meant to fixate on the mobilities of the film. While bicycles, motorcycles, and cars are connected to plot points, they are not mere tools as they are in *Drive*. The mobilities of the film are interwoven with every other aspect of Daniel's life to the point that they are taken in uncritically despite their critical importance to the greater context of the film. A point that I will return to, is that when speaking of landscape studies, Paul Groth has said of we Americans that not being able to see our landscape is like a fish not seeing the water.[58] I want to use what follows to amend Groth's claim. Not being able to see how we move through a landscape is akin to a fish not seeing the water. From an ideological perspective though, we must remember that we are not meant to see our mobilities; the water isn't supposed to be obvious to the fish. What happens if we start to pay attention to these mobilities?

Of course, we have been paying attention to our mobilities as the climate crisis grows more and more alarming. Fossil fuel burning machines are wrecking the climate and we must do something about this. The push to switch to electric cars appears to be finally gaining traction as I write this. But is that a good thing for the climate? Or must we tear down American automobility completely if we are to limit the ever-worsening sea-level rise, habitat loss, heat waves, and increase in hurricanes?

The real problem lies in the fact that despite the invisibility of our ideology of American automobility, our actions produce material consequences. And yet, it is that very invisibility that is preventing those in the wealthier nations of the world from doing meaningful change to cut emissions. If we are not even fully aware of how reliant we are on a driver-car assemblage to move through the world, how can we begin to fix that world?[59]

PLAN OF THE PRESENT WORK

My goal with *Ecomobilities* is to expand the Anthropocene's archive by questioning how American automobilities are represented in popular films. While that may seem like a rather academic effort, a bit of using a microscope when a telescope is demanded, it is my contention that this is the very type of work those trying to extract the meaning of the Anthropocene should be doing. We know very well that nature and society are linked and we know that this hybrid world around us is burning. Why are we letting it burn, if not adding even more fuel to the fire? I am convinced that the answer lies in ideology. Not in a vulgar, American political sense of red state versus blue state (those evil climate deniers!), but in a much more subtle manner in which everyday life is made natural to the point in which we cannot even begin to envision change. We know that fossil fuel emissions are contributing to increased greenhouse gases, but we do not know just how difficult it will be to give up our automobilities as a means to combat climate change. Chapter 1 is a sustained exposition on my method for those interested in that kind of thing (i.e., those who enjoy plunging into Deleuzian thought), and will outline my rational behind the use of the films that follow. In short, I am following Adorno and Horkheimer's treatment of popular film as a commodity despite the artistic intent of the filmmakers.

To begin the application of this new Anthropocene archive and the layering of place, humans, and machines, chapter 2 introduces the presence of automobility, ideology, and the Anthropocene in George Miller's *Mad Max: Fury Road.* The film notably does not take place in the United States, but nonetheless represents the American automobility response to a climate apocalypse. Despite the very fact that cars got us into this mess, we cannot move throughout the wasteland without one.

The film was made in the twenty-first century, well within our present moment of climate awareness, but it builds on a world first envisioned in the late twentieth century. The film's titular character, Max, inhabits an apocalyptic planet of harsh landscapes and violent encounters, but can only survive such a world through the use of machines, especially automobiles. While the film has been read as a feminist resistance to a patriarchal regime of fossil fuel dependency, I explore what the film has to say about the fusion of humans and machines within specific environmental conditions. Max's world reveals a specific ecomobility; one cannot survive without moving with machines. I will also discuss the degree of human/machine fusion. What are we Anthropocene-epoch travelers? Cyborgs? Hybrids? Assemblages? Perhaps all three? Ultimately, *Fury Road* shows that automobilities need not be the cause of environmental collapse. Unchecked capitalism has led to the apocalyptic hellscape. Automobiles, specifically aggressive SUVs, play a fundamental role in how humans experience nature in *Mad Max: Fury Road*. The film thus provides an entry into thinking about how Americans use cars as part of an assemblage or hybrid to move through environments while operating within a specific ideology.

Chapter 3: "Machines Precede the Climate: The Technological Fix" focuses on the concept of the Anthropocene and the magic bullet of technology. David Harvey draws attention to a footnote in the first volume of Capital that he argues is Marx conceiving of an assemblage between the modes of production, technology, social relations, nature, and ideas about the world.[60] I use this chapter to question our present assemblage while also utilizing Jean Baudrillard's concepts of simulacra and simulation to build upon Fisher's *Capitalist Realism*, again, the idea that we cannot move past capitalism. Fisher speaks of "the widespread sense that not only is capitalism the only viable political and economic system, but also that it is now impossible to even *imagine* a coherent alternative to it."[61] As Fisher points out, modern films about dystopian futures often fail to think past a capitalist system. Many of these dystopian films, I argue, use a framing that treats the Earth's atmosphere as "the desert of the real."[62] I use this chapter to focus on films that depict dystopian futures, but specifically address the idea of a technological fix in addressing the problems inherent in such a world. My argument is that for the most part, these types of films reflect a greater societal idea that the machines precede the climate, in that the very machines that got us into the climate crisis will somehow get us out of it. The machines have become more real than the atmospheric conditions. Rather than look at a single film like the previous chapter, I spend time with *I Am Mother*, *Pacific Rim*, and *Snowpiercer* to show how bleak futures are addressed through the production of new technology. The films present different technological fixes that have questionable successes in each of their dystopian futures and yet, we still hold out for such a fix to our own changing climate.

Chapter 4: "Zombies and the Horror of Not Having a Car: Apocalyptic Stories as Ecomobilities" addresses another kind of dystopian future. Where the previous chapter explored using machines and mobilities to address the Anthropocene, this one examines local responses to environmental disasters. Placed in the context of the COVID-19 pandemic, I explore how horror films have naturalized the need for American automobility when faced with an external threat. Further, specific ecomobilities emerge in which movement is connected to not just the landscape but themes of habitat, consanguinity, and masculinity. I first start with the zombies of *28 Days Later* which, despite their ontological threat in the film, allow director Danny Boyle to play with themes of familial relations and the putting down of roots. While never explicit in the film, American automobility is a commonsense component to surviving in this hellscape.

One need not have an explicit horror in the film to produce automobile dependent ecomobilities though. I also discuss the 2018 film *How It Ends*, which invokes both real and implied horrors that demand a cross-country road trip. A conservative masculinity is directly connected to the road trip and how one must deal with an external threat. This interweaving of automobility, masculinity, and end times shows how multiple ideological fantasies work in a Deleuzian sense to produce a very specific ontology of the near future.

Chapter 5: "I Hope You Have a Big Trunk 'Cause I'm Putting My Bike in It: Alternative Transportation as a Reinforcement of Capitalism" addresses the second alternative of which McKenzie Wark speaks, the idea that the only solution to climate change is to turn back time to a pre-capitalist society. This is not exclusive of failing to imagine an alternative to capitalism. I first discuss cycling as a resistance to American automobility both in practice and as depicted in films like the *40-Year-Old Virgin* and *Pee-Wee's Big Adventure* to explore how automobile alternatives are presented as deviant. I argue against the simple binary of cycling versus the automobile and instead push for a Žižekian understanding of resistance through a read of *L.A. Story*. I then explore the Pixar film *Onward*, which laments how scientific progress has eradicated the idea of magic, all the while using the automobility of a van to grasp that past sense of wonder. Even in trying to envision a simpler past, we cannot move past American automobility.

I then conclude this work and discuss how we move forward. What is most important in the connection between the automobile and the Anthropocene is the work being done by the assemblage of human, machine, and environment, as well as the ideological work at play. American automobility and capitalism are inextricably linked which begs the question as to whether we can, should, or must, envision the end of the personal car.

For those looking for a clear excoriation of the American automobile or a romantic look at a time when going for a drive had an innocence to it, I am sorry. Obviously, with what we know about fossil fuel emissions and their connection to the climate crisis, one cannot (or at least, should not) romanticize the automobile. And yet, I will admit that I find myself unable to completely eschew such a form of mobility. Whether I am out working on my Jeep, talking to one of my kids about learning to drive, or enjoying a film like *Ford v. Ferrari*, I am in American automobility's grasp. Its ideological hold is strong, and these days in the midst of the COVID-19 pandemic, I find it easier to imagine the end of the world than the end of driving a car.

NOTES

1. World Meteorological Organization, "New climate predictions increase likelihood of temporarily reaching 1.5°C in next 5 years," May 27, 2021, https://public.wmo.int/en/media/press-release/new-climate-predictions-increase-likelihood-of-temporarily-reaching-15-°c-next-5.

2. Intergovernmental Panel on Climate Change, "Summary for Policy Makers," in *Global Warming of 1.5°C. An IPCC Special Report on the impacts of global warming of 1.5°C above pre-industrial levels and related global greenhouse gas emission pathways, in the context of strengthening the global response to the threat of climate change, sustainable development, and efforts to eradicate poverty*, edited by V. Masson-Delmotte, P. Zhai, H.-O. Pörtner, D. Roberts, J. Skea, P.R. Shukla, A. Pirani, W. Moufouma-Okia, C. Péan, R. Pidcock, S. Connors, J.B.R. Matthews, Y. Chen, X. Zhou, M.I. Gomis, E. Lonnoy, T. Maycock, M. Tignor, and T. Waterfield, https://www.ipcc.ch/site/assets/uploads/sites/2/2019/05/SR15_SPM_version_report_HR.pdf, 7.

3. Rebecca Hersher, "Earth is Barreling Toward 1.5 Degrees Celsius of Warming, Scientists Warn," *National Public Radio*, May 26, 2021. https://www.npr.org/2021/05/26/1000465487/earth-is-barreling-toward-1-5-degrees-celsius-of-warming-scientists-warn; Zander S. Venter, Kristin Aunan, Sourangsu Chowdhury, and Jos Leleiveld, "COVID-19 Lockdowns Cause Global Air Pollution Declines," *Proceedings of the National Academy of Sciences* 117, no. 32 (August 11, 2020): 18984–18990.

4. Richard Grove, *Green Imperialism: Colonial Expansion, Tropical Island Edens, and the Origins of Environmentalism, 1600-1860* (Cambridge: Cambridge University Press, 1995), 166.

5. Spencer R. Weart, *The Discovery of Global Warming, Revised and Expanded Edition* (Cambridge: Harvard University Press, 2008), 2–3.

6. Weart, *Discovery*, 7.

7. Paul J. Crutzen and Eugene F. Stoermer, "The 'Anthropocene'," *Global Change Newsletter*, no. 41 (2000): 17.

8. One exception comes from William Ruddiman who argues for significant anthropogenic climate change happening roughly 5,000 years ago with advances in agriculture. William F. Ruddiman, "The Anthropogenic Greenhouse Era Began Thousands of Years Ago," *Climatic Change*, no. 61 (2003): 261–293.

9. Intergovernmental Panel on Climate Change (IPCC), "FAQ Chapter 1," https://www.ipcc.ch/sr15/faq/faq-chapter-1/.

10. Andreas Malm, *Fossil Capital: The Rise of Steam Power and the Roots of Global Warming* (Lodon: Verso, 2016), 391.

11. Malm, Fossil, 391.

12. Malm, Fossil, 391, his emphasis.

13. Jason W. Moore, *Capitalism in the Web of Life: Ecology and the Accumulation of Capital* (London: Verso: 2015), 4.

14. Moore, *Capitalism*, 30, his emphasis.

15. Moore, *Capitalism*, 291, his emphasis.

16. Neil Smith, *Uneven Development: Nature, Capital, and the Production of Space* (Athens: University of Georgia Press, 2008), 68.

17. Karl Marx, *Capital: A Critique of Political Economy, Volume One* (London: Penguin Books, 1990 [1867]), 273.

18. Marx, *Capital Vol 1*, 290.

19. Moore, *Capitalism*, 303.

20. Jason W. Moore "The Capitalocene, Part I: On the Nature and Origins of Our Ecological Crisis," *The Journal of Peasant Studies* 44, no. 3 (2017): 596.

21. Moore, *Capitalism*, 303.
22. Joel Wainwright and Geoff Mann, *Climate Leviathan: A Political Theory of Our Planetary Future* (London: Verso, 2018), ix.
23. Bruno Latour, *Facing Gaia: Eight Lectures on the New Climatic Regime* (Cambridge: Polity Press, 2017), 122.
24. Mark Fisher, *Capitalist Realism: Is There No Alternative?* (Winchester, UK: Zero Books, 2009).
25. McKenzie Wark, *Molecular Red: Theory for the Anthropocene* (London: Verso, 2016), xxi.
26. Wark, *Molecular*, xxi.
27. Jedediah Purdy, *After Nature: A Politics for the Anthropocene* (Cambridge: Harvard University Press, 2015), 5.
28. Moore, *Capitalism*; Smith, *Uneven*; Latour, *Facing Gaia*; Donna J. Haraway, *Staying with the Trouble: Making Kin in the Chthulucene* (Durham: Duke University Press, 2016).
29. Purdy, *After Nature*, 15.
30. Althusser, "Ideology," 120.
31. Smith, *Uneven*, 28.
32. Marx and Engels, "The German Ideology," 154–155.
33. Marx and Engels, "The German Ideology," 172, their emphasis.
34. Althusser, "Ideology," 125.
35. Althusser, "Ideology," 122, his emphasis.
36. Althusser, "Ideology," 126.
37. Althusser, "Ideology," 128.
38. Althusser, "Ideology," 135.
39. Stuart Hall, "Signification, Representation, Ideology: Althusser and the Post-Structuralist Debates," *Critical Studies in Mass Communication* 2, no. 2 (June 1985): 100, his emphasis.
40. Hall, "Signification," 105.
41. Slavoj Žižek, *The Sublime Object of Ideology* (London: Verso, 2008 [1989]), 25.
42. Žižek, *Sublime*, 30, his emphasis.
43. Cotten Sieler, *Republic of Drivers: A Cultural History of Automobility in America* (Chicago: The University of Chicago Press, 2008), 7.
44. Seiler, *Republic*, 147.
45. Ann Laura Stoler, *Duress: Imperial Durabilities in Our Times* (Durham: Duke University Press, 2016), 110.
46. Seiler, *Republic*, 2.
47. Seiler, *Republic*, 29.
48. Seiler, *Republic*, 90.
49. Seiler *Republic*, 6.
50. Seiler, *Republic*, 58.
51. Seiler, *Republic*, 128.
52. Seiler, *Republic*, 10.

53. Michel Foucault, *Discipline and Punish: The Birth of the Prison* (New York: Vintage, 1995 [1975]), 23–24.

54. Seiler, *Republic*, 35.

55. Michel Foucault, "Truth and Power," in *Power/Knowledge: Selected Interviews & Other Writings 1972-1977*, edited by Colin Gordon (New York: Vintage Books, 1980 [1977], 118.

56. Althusser, 120.

57. Full disclosure, I first saw the film with my dojo. I was six and ditched my mom to sit next to my sensei. The context of seeing a film is important to ideological reproduction.

58. Paul Groth, "Frameworks for Cultural Landscape Study," in *Understanding Ordinary Landscapes*, edited by Paul Groth and Todd W. Bressi (New Haven: Yale University Press, 1997), 1.

59. Tim Dant, "The Driver-car," *Theory, Culture, & Society* 21, no. 4–5 (2004): 61–79.

60. David Harvey, *A Companion to Marx's Capital: The Complete Edition* (London: Verso: 2018 [2010]), 191–214; Karl Marx, *Capital: A Critique of Political Economy, Volume One* (London: Penguin Books, 1990 [1867]), 493–494.

61. Fisher, *Capitalist Realism*, 2, his emphasis.

62. Slavoj Žižek, *Welcome to the Desert of the Real* (London: Verso, 2002), 15.

Chapter 1

Method

If the work at hand is an effort in uncovering ideology, the archive cannot be confined to scientific studies. Nor is this a work of knowledge production in the power/knowledge sense.[1] Instead, to grasp the Gramscian naturalization of ideas, that is, the ways in which the automobile has been made an unquestionable part of society in the Anthropocene, we must turn to media in some form.

New technology produces new art. Obviously. The camera brought forth a new medium upon which artists could work. But new technologies go beyond the technical aspect of art and in fact shape the very conditions of possibility for the stories that can be told through that art. For example, those of a certain age can remember when access to a telephone made for dramatic cinematic storytelling. Telephones were connected to fixed places meaning that any long distance yet real-time communication between characters was limited to specific locations. Further, the severing of actual wires would render the telephone useless and thus heighten the tension. In the 1999 film *The Matrix*, human characters pass in and out of the "Matrix," the elaborate software that most humans view as reality, through hardwired telephones. The film was noted for its special effects and its ability to make a dystopian future seem plausible but viewing it twenty years later shows how it is a product of its time. The film opens with Trinity (Carrie Ann Moss), one of the humans who understands that the Matrix is not reality, trying to evade the Agents (led by Hugo Weaving), which are anthropomorphized algorithms that are working to stop the humans who have knowledge of the Matrix. After a lot of cool gunfire and dramatic chase violence, Trinity runs to a phone booth, a set location where she can use the phone to exit the Matrix and get back to the flying motor home thing on which she and her crew live. The action is driven by the need to get to a specific place.

Mobile phones, of course, dramatically changed how these tense scenes play out. While the characters in *The Matrix* had access to mobile phones, they were not connected to Wi-Fi. They could make calls, but presumably entering and exiting the Matrix required a dial-up modem. This made specific locations with working wired telephones crucial to the protagonists' success. Tense cinematic stories are now told within the social fact that long distance communication through a call or text message is almost aspatial. Apart from remote wilderness locations or authoritarian nations, mobile phones are easily used. Plots must advance with new conventions to evoke tension. For example, in the 2018 film *The Commuter*, Liam Neeson plays a recently laid-off insurance salesman, Michael MacCauley. On his commute home from the city to the suburbs, just before he boards the train, his cellphone is taken, presumably having been stolen by the stranger who bumped into him on the platform. The train is a confined space, "a travelling incarceration" as de Certeau puts it.[2] Michael is now trapped within the train until he reaches his stop and has no means with which to speak with his family. Other train passengers cryptically speak on their phones, showing their connection to the world outside of the train as well as giving the audience clues to who they are. Even the types of phones are a character trait; we know that a Wall Street day trader (Shazad Latif) is obnoxious precisely because he is loudly speaking into a Bluetooth headset. The choppy montage of shots from within the hot, crowded train evokes confinement as well as alerts the audience to the impending trouble (it *is* a Liam Neeson movie after all) making us suspicious as to the passengers' true motives. A mysterious woman, Joanna (Vera Farmiga) approaches Michael offering him a large sum of money if he simply can identify one of the other passengers on the train who goes by the alias of "Prynne." It becomes clear that Prynne will be killed, but Michael appears to have no choice but to go along. Michael borrows a cellphone from another passenger to try and get help. While the mobile nature of the phone seemingly frees the communication, the phone has no signal while the train is underground. He is eventually able to make a call and leave a message with his friend in the police department, but he is then called on the phone by the mysterious Joanna. She threatens Michael's family's safety if he tries to contact the police, forcing him to focus on finding the passenger.

Tension is still evoked through communication, or the impossibility of that communication, but the changing nature of technology changes how the story unfolds. Since cellphones seemingly make telecommunication effortless, Michael must lose his phone for the plot to be remotely plausible. The ever-present phone network also makes the seemingly omniscience of Joanna possible. If phones communicate via invisible waves, why couldn't the bad guys easily monitor mobile phones? Despite the apparent aspatial nature mobile phones, the film shifts the spatial tension to the confined train cars and the fact that Prynne will be departing the train at a specific station. The tension is not much different from a twentieth-century thriller but must be told in a new way in order for the audience to accept it.

While communication technologies have changed dramatically in a short amount of time, the stories we tell through mobilities are strikingly resilient. The first automobile was patented by Carl Benz in 1886 and notably the first film was made in 1888 by Louis le Prince. Today, both cars and films have independently evolved significantly while keeping a few vestigial traces of their origins. And yet, there has been little change in the stories told with, by, and for the automobile in film. In a rather Gramscian sense, the automobile is now a natural component of mobility in the developed world.[3] Even when we are critiquing our dependence on automobiles, we cannot think outside of an ideology of American automobility. Further, the durability of this form of automobility suggests that a radical restructuring of American daily life to reduce emissions and combat climate change will not be easy. My goal in using film here is to simultaneously build on our understanding of American automobility as well as to make an argument as to why the myriad work alerting us to the dangers of climate change has elicited so little response.

EXPANDING THE ANTHROPOCENE'S ARCHIVE

In terms of an archive, I am not looking to art house films, nor looking at films that were highly successful or highly praised. I am interested in mass culture that is easily streamed and consumed within one's home. I use the term mass culture while nodding to Adorno and Horkheimer. I am not fully decrying the "culture industry" as the two did in their famous essay, nor am I critiquing art against fascism.[4] I do, however, think the mass production of films for theatrical distribution and streaming services is of value to my study precisely because of the sameness critiqued by Adorno and Horkheimer. They saw mass culture as being far from the creation of art and rather an effort to produce a commodity no different from a home appliance or automobile:

> Even the differences between the more expensive and cheaper models put out by the same firm steadily diminish: for automobiles, there are such differences as the number of cylinders, cubic capacity, details of patented gadgets; and for films there are the number of stars, the extravagant use of technology, labor, and equipment, and the introduction of the latest psychological formulas.[5]

If we do not get hung up on the artistic merits of a film and follow Adorno and Horkheimer in arguing that such mass culture is a commodity, we can extract ideologies at work; not just capitalism, but as I will show, a very specific automobility that is uncritically reproduced in mass culture films.

Some of the films mentioned in what follows will be the obvious profit-driven movies that one cannot help but eat popcorn to, but others might suggest a more artistic production. In keeping with Adorno and Horkheimer, the difference is irrelevant here. Films that push against the mass produced feel of Hollywood might be more entertaining to watch, but they are not a resistance to mass culture. "Whenever Orson Welles offends against the tricks of the trade, he is forgiven because his departures from the norm are regarded as calculated mutations which serve all the more strongly to confirm the validity of the system."[6] This is a rather Žižekian read (Or rather, Žižek owes much to the Frankfurt School) in that power has already accounted for resistance.[7] This same concept works as we focus on the ideology of American automobility itself. Even those films that appear to be offending the system are often reproducing it.

I also will borrow from Deleuze and, at times, Guattari in my readings of the films that follow. I find Deleuze's work to be inherently spatial, even when he is focused on time, and of great use in talking of both mobilities and cinema. In his own philosophies of film, Deleuze discusses classic techniques of *montage*, the stitching together of individual shots to produce a specific affect or convey an idea, but does so to get at new perceptions of space and time. He invokes Henri Bergson right away to lay groundwork for concepts like time and movement. Deleuze pulls out three theses on movement from Bergson's *oeuvre*, the first being that movement is "distinct from the space covered."[8] This, Deleuze contends, is more complex than it might first seem. Movement is an indivisible concept; it can only occur through the succession of spaces covered. A film, with its stitching together of images, produces "the movement-image."[9] Bergson's second thesis reveals an ambiguity of art and science present in cinema. Classical thought saw movement as occurring between rather formal, ideal poses, whereas modern science introduced the approach of deriving movement from separate, observed elements. Bergson, again according to Deleuze, saw cinema as "the organ for perfecting the new reality."[10] And finally, and what I see as most important, is Bergson's third thesis of movement and change. In mobilities studies, we typically use the simple A → B to designate mobilities, with the arrow being the important thing in this signifier.[11] What Bergson does, however, is challenge the simplicity of this vector. Deleuze puts it this way:

> If I consider parts or place abstractly—A and B—I cannot understand the movement which goes from one to the other. But imagine I am starving at A, and at B there is something to eat. When I have reached B and had something to eat, what

> has changed is not only my state, but the state of the whole which encompassed B, A, and all that was between them.[12]

The key to Bergson, and Deleuzian Bergsonism, is the qualitative nature of an event and that the whole changes, though not in terms reducible to succession. To map out movement upon space is to eradicate the inherently qualitative and internal nature of mobilities. Suzanne Guerlac sums up Bergson's thoughts on mobility succinctly: "Mobility is not a thing, it is an action. As such, it is indivisible. It cannot be divided up into units, counted, or mapped onto space."[13] The internal quality of movement is key to Bergson's concept of pure duration, which he defined as "the form which the succession of our conscious states assumes when our ego lets itself *live*, when it refrains from separating its present state from its former states."[14] Thinking in space, for Bergson, is all about the separation and mapping out of one moment to the next, which ruins any conception of the whole. Thinking in duration, on the other hand, is to conceive of "nothing but a succession of qualitative changes, which melt into and permeate one another, without precise outlines, without any tendency to externalize themselves in relation to one another, without any affiliation with number: it would be pure heterogeneity."[15]

Right away, my claim to the spatiality of Deleuze might appear to be challenged by his very invocation of Bergson in which he argues, "state problems and solve them in terms of time rather than of space."[16] What Bergson was attacking was the treating of space as simply an empty canvas upon which events could be set, which he saw as stripping those events of any qualitative meaning. Bergsonian duration, "a way of being in time," and the importance of multiplicities, is perhaps best shown as an equally spatial concept by Doreen Massey.[17] It is important to remember that Bergson is critiquing a mathematical conception of space and not the spatial turn.[18] What Bergson was working toward was not an elimination of space, but a reconfiguring: "Space itself will need to be based in things, in relations between things and between durations, to belong itself to the absolute, to have its own 'purity.'"[19] This is exactly what Massey gets at with her theory of space, often invoking both Deleuze and Bergson, to argue for space as being "a simultaneity of stories-so-far."[20]

I will continue to insist on a conception of mobilities that is a connection between the automobile, the crossing of physical spaces, the production of ecosystems, and the formation of subjects that is in the vein of Bergson, Deleuze, and Massey. The problem, unfortunately, is the difficulty in representing pure duration through writing without turning it into Bergson's definition of space.[21] To use film as an archive is to do double work when it comes to mobilities studies and to work toward revealing ideology's hold. On the one hand, the experience we have in watching a film, much like Bergson's example of music, is a succession of moments in conjunction with the viewer that cannot be split into individual objects. "[I]f we interrupt the rhythm by dwelling longer than is right on one note of the tune, it is not its exaggerated length, as length, which will warn us of our mistake, but the qualitative change thereby caused in the whole of the musical phrase."[22] The film, as a creative endeavor that relies on interpretation from its viewer, allows for a study of duration along these lines. The montage of scenes, along with layered soundscapes, and even the conditions in which the final product is consumed, produce a whole to be experienced rather than mapped out as a succession of events. I will provide such a succession of events as I explain key moments to argue specific points, but my ultimate goal is to evoke the lived experience of viewing the film to elucidate the role of American automobility in everyday life. This ties in with the double benefit of using film as an archive, in that film allows for the naturalization of ideology through its very montage of elements. As I will work to show throughout what follows, the experience of viewing a film allows for Gramscian "common sense" to be produced and reinforced, especially when it comes to the use of a car.[23]

IMAGINING THE END OF THE WORLD

A recurring theme in what follows is the notion of the end of the world, at least as we know it. I will keep referencing Mark Fisher's *Capitalist Realism* in our inability to imagine the end of a capitalist world.[24] As many of the films mentioned throughout will show, there is no struggle to present an apocalyptic scenario that represents the anxiety and looming disaster of climate change. And yet, those stories so often fail to address, let alone think past the structures of capitalism. Automobility is fully embedded within capitalism, as both a commodity to desire and a means to reproduce the relations of power that keep it running.

The prevalence of capitalism in bringing about the Anthropocene/Capitalocene demands a materialist reading of the films that follow, despite the overwhelming attention to ideas. This is what makes these films such ripe sources of discovery about what is deemed normal in this new epoch. Every film here is dealing with some amount of anxiety, most of it environmental in form. The films are also big and fantastical, sometime absurd. The spectacle is precisely why we watch these types of movies, but what happens when we methodically peel back the layers of fantasy like an archaeologist uncovering a site? We are left with the material world, represented in full on the big screen or our televisions. The trick will be to not just expose the material reality, but to grab a hold of it, change it as necessary, and work toward a more just and equal epoch.

NOTES

1. Michel Foucault, *The Archaeology of Knowledge and the Discourse on Language*, Translated by A.M. Sheridan-Smith (New York: Pantheon Books, 1972 [1969]), 27.

2. Michel de Certeau, *The Practice of Everyday Life* (Berkeley: University of California Press, 1988), 111. This concept of the train as a space of disciplinary power will be explored more in chapter 2.

3. Antonio Gramsci, *Selections from the Prison Notebooks*, edited and translated by Quintin Hoare and Geoffrey Nowell Smith (New York: International Publishers, 1971), 375–377.

4. Theodor Adorno and Max Horkheimer, "The Culture Industry: Enlightenment as Mass Deception," in *Dialectic of Enlightenment* (London: Verso, 1997 [1944]), 120–167.

5. Adorno and Horkheimer, "Culture Industry," 123–124.

6. Adorno and Horkheimer, "Culture Industry," 129.

7. Slavoj Žižek, *The Plague of Fantasies* (London: Verso, 2008 [1997]), 31.

8. Gilles Deleuze, *Cinema 1: The Movement-Image* (Minneapolis: University of Minnesota Press, 1986 [1983]), 1.

9. Deleuze, *Cinema 1*, 3.

10. Deleuze, *Cinema 1*, 8.

11. Tim Cresswell, *On the Move: Mobility in the Modern Western World* (New York: Routledge, 2006), 2–4.

12. Deleuze, *Cinema 1*, 8.

13. Suzanne Guerlac, *Thinking in Time: An Introduction to Henri Bergson* (Ithaca: Cornell University Press, 2017), 67.

14. Henri Bergson, *Time and Free Will: An Essay on the Immediate Data of Consciousness* (Mineola: Dover, 2001 [1913]), 100, his emphasis.

15. Bergson, *Time and Free Will*, 104.

16. Gilles Deleuze, *Bergsonism* (New York: Zone Books, 2006 [1966]), 31.

17. Doreen Massey, *For Space* (London: Sage Publications, 2005).

18. Henri Lefebvre, *The Production of Space*, translated by Donald Nicholson-Smith (Malden: Blackwell Publishing 1991 [1974]); David Harvey, *The Condition of Postmodernity* (Malden: Blackwell Publishing, 1990), 201–323; Massey, *For Space*.

19. Deleuze, *Bergsonism*, 49.

20. Massey, *For Space*, 9, 20–24.

21. Deleuze, *Bergsonism*, 22.

22. Bergson, *Time and Free Will*, 100–101.

23. Gramsci, *Prison Notebooks*, 348.

24. Mark Fisher, *Capitalist Realism: Is There No Alternative?* (Winchester, UK: Zero Books, 2009).

Chapter 2

"So Shiny So Chrome"

Images and Ideology of Humans, Machines, and the Earth in George Miller's Mad Max: Fury Road

George Miller's 2015 film, *Mad Max: Fury Road*, is violent, fast, loud—furious, I suppose sums it up best—yet it still manages to demand interpretations of its noticeably spartan and linear storytelling. The film is so linear, in fact, that it consists of a drive out and back; the characters do not even loop their journey to cover new ground. Wives escape from a warlord, they turn around, and they take over the warlord's compound, the end. The characters are like sharks; they move to remain alive. Critiques of the film see it as a feminist action film, a feminist failure due to its excessive violence, masculinity in crisis, a biblical exodus, and an environmental warning.[1] Yet, none of these readings of the film take into account the interplay of humans, machines, and the Earth in *Fury Road*. Can anyone sincerely argue that the vehicles of *Fury Road* are less important to the film than the women? It is clear from the outset that Miller is blaming a modern dependence on oil for the nuclear wasteland in which the characters inhabit, but he falls short of suggesting a resistance to the automobile. In fact, to read *Fury Road* as a resistance to a patriarchal system of machinic mobility is too simple. This film is a celebration of machines that are imbricated into ideologies of capitalism, while it simultaneously critiques the violence of capitalism that leads to such an apocalyptic wasteland. *Fury Road* reveals an ideology of ecomobility, that is, a way in which human bodies ought to move through the Earth's ecosystems. This specific ecomobility blends human and machine into the only assemblage/hybrid capable of travelling the wild and desolate spaces of this apocalyptic wasteland. Miller has made a film that reveals the complexity of automobility, a system that is both positive and negative in its effects upon the Earth.

In what follows, I show the interplay of automobility and ecomobility in *Fury Road*. This chapter proceeds in four parts. First, Max, the film's titular character, is introduced to provide context for the film as well as my reading of it. I then explore the ideology of automobility within which the film was made and use Žižek's ideas on ideological distancing to explore automobility's durability. Next, I explore the connection between humans and automobile and show how *Fury Road* complicates prevailing notions of human/machine assemblages in automobility studies. I call for a greater focus on hybridity and the complicated relationship between humans, machines, and the environment. Finally, I argue that fixating on the over-the-top vehicles of *Fury Road*, like fixating on the evils of Sport Utility Vehicles, masks the underlying violence of capitalism itself. The violence of the men and women of the wasteland are the images upon which we are meant to fixate in the film, but such images actually get in the way of a pure critique of automobility. Rather, Miller's critique of the underlying capitalist structures of automobility only peeks through the wild car chases and explosions. It is the ideological violence of capitalism driving our systems of mobility that led to Max's world, not simply the internal combustion engine. This over-the-top action film provides a means to questioning the ideological distancing at work in automobility.

When I use the concept of automobility, I am invoking John Urry's playful definition:

> On the one hand, "auto" refers reflexively to the humanist self, such as the meaning of "auto" in autobiography or autoerotic. On the other hand, "auto" refers to objects or machines that possess a capacity for movement, as expressed by automatic, automaton, and especially automobile. This double resonance of "auto" is suggestive of how the car-driver is a "hybrid" assemblage, not simply of autonomous humans but simultaneously of machines, roads, buildings, signs and entire cultures of mobility.[2]

Urry sees automobility as not only about the automobile, though driving one is clearly a part of the system of automobility. Further, these machines, at least so far as the technology exists at the time of this writing, must fuse with humans in order to move and are "only able to roam in certain time-space scapes."[3] Humans and automobiles assemble in a specific time and space; this type of mobility demands it.

Böhm et al. make the claim that automobility is impossible to maintain. Automobility is three things: an important, modern institution, an ideology, and a specific phenomenological way of experiencing the world.[4] It is so pervasive and power-laden that Böhm et al. use "regime" in place of "system" to fully describe how this specific mobility impacts contemporary society. Echoing Foucault's "where there is power, there is resistance," Böhm et al. argue that automobility has been historically and recently contested and the question is, how has such a regime maintained its grasp of power?[5] Their answer is equally Foucauldian:

> The attribution of deviance to alternatives to the car means that those advocating such alternatives have trouble articulating successfully their own regime of truth regarding cars—we don't believe their statements of "fact" because they are already regarded as deviant.[6]

Power relations are so layered into mobility practices that subjects are not free to move through spaces without the aid of the automobile. But, is it as simple as that? Does the modern subject not ride a bicycle to the store because of the practice's deviance? There are clearly negative endings presented by the automobile (car accidents, climate change, oil dependency, resource-based wars) that are accepted as "fact" by drivers. And yet, modern humans struggle to envision a mobility free of internal combustion driven machines. Where Böhm et al. see regimes of truth producing a society in which self-propelled machines are the only way in which to move through space, that very *space* is overlooked. A focus on ecomobility, exploring how space and mobility operate in dialectical tension rather than passive and active roles, respectively, as well as a deeper study of ideology will open new possibilities for automobility studies. Those certain "time-space scapes" of which Urry speaks should not be thought of as abstract spatio-temporal planes, but rather material ecosystems through which humans move. Paul Groth has written that humans not being able to read the landscape is "like fish that can't see the water."[7] Urban planners will look to the city as a cure for automobility and its environmental problems, but mobility outside of dense urban living is something entirely different.[8] We must also look at the un(der)developed spaces through which humans move. There are spaces of the world that the bicycle is not selected to help travel through, not because of its deviance, but because it is not the proper machine with which to assemble. The proper assemblage depends on the landscape.

It may seem odd to use a bizarre narrative film to draw attention to the material ecologies through which automobility roams and its resulting ideology, but Miller's film lays bare the seemingly absurd nature of automobility. This automobility regime is presented absurdly, which allows the underlying workings of automobility to briefly become visible for the film's spectators. It is worth considering that if the characters of *Fury Road* cannot remove the ideological framings of automobility to construct a world free of a dependence on the internal combustion engine, what hope do the rest of us have?

IGNITION: MEETING MAX

The film's prologue opens with nothing but Max Rockatansky's (Tom Hardy) voice stating, "My world is fire and blood." His gravelly voice is replaced by a cacophony of news reports discussing first oil and then water wars being waged due to the lack of these important resources. While the former is necessary for the biological continuation of life, the latter is only necessary for powerful machine-assisted mobility. Immediately, Miller has tipped his ideological hand. Water and oil hold the same weight and the haunting sounds layered upon a dark screen focus the viewer on this fact. Max's world maybe fire and blood, but that is only due to the fact that this world is lacking water and oil.

The voices shift from speaking of geopolitical news reports and commentary to comparing this new life to disease and decay. "The Earth is sour," says a woman's voice. "Our bones are poisoned, speaks another. "We have become half-life," claims a third. Max resumes his monologue, "it was hard to know who was more crazy . . . me, or everyone else." The black screen cuts to Max pissing on barren red soil, his heavily modified Ford Falcon V-8 Interceptor parked next to him. Max and machine are framed by two rugged buttes with a desolate valley stretching out to the horizon. The camera pedestals down to a two-headed lizard, which we are to assume is the result of nuclear war. The lizard scurries off toward Max, who without turning squishes it with his boot heel. He quickly grabs it and shoves the lizard into his mouth, noisily chewing. What first appears as animal cruelty is clearly a matter of survival in the desert.

The lizard consumed, Max moves into action. There is a montage of Max picking up his bedroll, assembling his few belongings and getting into his car. Far from subtle, the quick, discontinuous cuts are layered with the sounds of engines, not just revving, but moving near and far. Max himself is not moving much, but the soundscape matched with images of the car and the gear necessary to live in this wilderness produce a space of ideological ecomobility. Miller's aural and visual montage reinforces the viewer's ideological spectacles of an oil-driven ecomobility.[9] Max's existence in this wilderness demands a gasoline-burning vehicle in addition to his rucksack full of utensils and a bedroll. An assemblage *must* occur. Miller's layering of image and sound not only develop this idea, but the montage produces mobility even where there is no movement. Eisenstein's "association montage" is at play, in which the montage of shots produces a dynamic subject, "not in the field of space but of psychology."[10] The viewer understands the movement mentally if he or she is within the ideology of automobility.

Max speeds off into the valley below; his dust dissipates as does the sound of his car. This is the same framing from the opening scene, presenting the viewer with the desert wilderness free of humans and machines. This natural space is now quiet and looks untouched. What should be coded as peaceful nature is discomforting. Our unease comes from nature's stillness. Fortunately, the calm is broken by motorcycles and trucks driving into the shot, angrily chasing Max. The machines belong in this space.

Max is quickly captured and his voice tells us he is "a man reduced to a single instinct . . . survive." Survival means operating within a specific eco-mobility in this wilderness.

IDLING: IDEOLOGIES OF AUTOMOBILITY

George Miller made a film about a possible near future, but he was working within a definite time and space, specifically at the beginning of the Anthropocene, the proposed geologic point in which humans got into the driver's seat of the Earth's climate. While the Anthropocene starts in the eighteenth century with the beginning of significant fossil fuel use, the concept is a twenty-first century one.[11] The warming, changing climate is clearly happening due to human causes, hence the epoch's name, and the automobile has been held as a major culprit in adding carbon dioxide to the atmosphere and increasing our greenhouse gases.[12] If oil wars don't lead to Miller's nuclear wasteland, our oil dependence threatens to destroy Earth regardless.

Yet, the call for sustainable transportation continues unheeded. The automobile has undergone technological changes of more efficient fuel consumption, electric motors, and so on, but even such "green" progress does not stop the system of automobility. In fact, those very green improvements designed to reduce the environmental effects of the automobile actually increase fuel consumption. Yes, efficient hybrid cars use less fuel than a comparably sized nonhybrid, but the technology leads to driving more miles at the same cost rather than conservation of fuel.[13] What this tells us it that automobility is not about rational decisions, nor is it a simple regime of truth. Urry refers to "cultures of mobility,"[14] but what happens if we instead focus on Böhm et al.'s ideologies of mobility? While ideology can evoke something like "false consciousness," that is, subjects are too dumb to know what they do, a more thoughtful analysis may prove useful. Slavoj Žižek has worked to rehabilitate ideology, setting it alongside Lacanian psychoanalysis and Hegelian dialectics. For Žižek, a key part of ideology is that it relies on fantasy to prevent its collapse. First, in his concept of the "ideological fantasy," Žižek states that ideology works on a double illusion. It is not that subjects do not know what they do, rather "they know very well how things really are, but still they are doing it as if they did not know."[15] The subject overlooks the illusion that structures reality and does so within the social field. It is not the case that other forms of automobility are surreptitiously made deviant; we can see through this, but we choose to maintain such an illusion. Even if we distance ourselves from the negative effects of automobility by purchasing a Prius, we are still fully, if not more, enmeshed in automobility. In fact, the act of buying a car with an electric motor conforms to Žižek's "fetishistic disavowal" that maintains a distance from a pure conformity to ideology to allow for the continuation of that ideology.[16] He uses the example of Kubrick's *Full Metal Jacket*. Vincent D'Onofrio's character fully conforms to discipline and ideology of the military, which ends in murder that does not advance the military's goals. Mathew Modine's character, however, maintains an ironic distance from the ideology, which allows him to be fully hailed by the ideological power. "He is the one in whom the interpellation by the military big Other has fully succeeded; he is the fully constituted military subject."[17] If we take automobility to be an ideology in the same vein, how would a hybrid vehicle accomplish the fetishistic disavowal? Böhm et al. signal automobility's environmental impossibilities:

> A second antagonism, which seems well established and understood today, points to the concerns about ecological sustainability of the contemporary regime of automobility . . . It contributes significantly to the depletion of non-renewable resources, notably oil (including production of plastics), rubber, platinum, lead, aluminum and iron.[18]

This depletion of nonrenewable resources is the impetus for the hybrid automobile. By reducing the oil consumed, one can resist the regime of automobility. But, such a distancing from the willful waste of oil is in fact a way to overlook the illusion of the ideological fantasy of automobility. "Conserving" fuel while not reducing mobility reinforces automobility's call to the subject. This is not resistance at all but a strengthening of this regime of power. Automobility is certainly not resisted by better fuel efficiency; wasteful automobility continues through the effort to curb fuel consumption and emissions that allows more cars to be sold and more fuel to be burned each year.[19] Not only are oil and iron consumed, but new minerals are mined for the new batteries now needed. Perhaps the answer to why automobility persists is to be found in our willingness to distance ourselves from it.

REVVING: HYBRID HUMANS AND MACHINES

Max's car is more than a generic object. The V-8 Interceptor connects *Fury Road* to the previous three Mad Max films as well as establishes the complex nature of human and machine. The car is connected to Max, but not fused to him. The term hybrid does not fully explain the relationship, as it indicates an offspring from the mating of human and machine. Tim Dant suggests in its place "assemblage."

> The particular driver-car may be assembled from different components with consequent variations in ways of acting, and its modal form may vary over time and place. However, despite variations, the assemblage of the driver-car enables a form of social action that has become routine and habitual, affecting many aspects of late modern society. . . . Neither the human driver nor the car acting apart could bring about the types of action that the assemblage can; it is the particular ways in which their capacities are brought together that bring about the impact of the automobile on modern societies.[20]

Max and his V-8 Interceptor can separate, which happens quickly in *Fury Road.* Another component of this driver-car assemblage is that Max can assemble with other machines throughout the film when the need arises. Max needs a machine to survive in this apocalyptic landscape, for, as Dant argues above, "neither the human driver nor the car acting apart could bring about the types of action that the assemblage can." This assemblage is necessary for specific practice, which ultimately means survival in this wilderness. One cannot exist in this space without such an assemblage. Further, the film does not present practices or ideologies in place of automobility. The breeding harem that escapes from Immortan Joe (Hugh Keays-Byrne) does so in a large truck. The women living in "The Green Place" use motorcycles. No one operates outside of the ideology of automobility; no one has a practice completely free of an assemblage between driver and car.

The use of assemblage evokes Deleuze and Guattari, though Dant states that similarity of the term to their "machinic assemblage" is "coincidental."[21] Yet, I would argue that both assemblages should be set against one another to fully understand the messiness of bodies and machines. Deleuze and Guattari offer the machinic assemblage as a conception of bodies interacting with other bodies, while an "assemblage of enunciation" refers to written and spoken language. These concepts connect to their larger project of linguistics and the relationship between sign and signifier, but these ideas are not that different from Dant's concept of the driver-car. Deleuze and Guattari describe a tetravalent assemblage, a four-part bonding of connections that comprise the greater machinic assemblage. The Earth is connected to social groups and there are class/status connections, but, using feudal Europe as an example, they also connect "the body of the knight and the horse to their new relation to the stirrup" as well as "the weapons and tools assuring a symbiosis of bodies."[22] Deleuze and Guattari are connecting human and tool as a symbiosis, an ecological term describing organisms working together, but not necessarily fusing into a hybrid. This machinic assemblage is but one side; mobility, or "deterritorialization," is also important for assemblages in general. Deleuze and Guattari hold the Crusades as an example. The knight and horse assemble, but they are also moving to the East, which cannot be ignored.[23] Deterritorialization is a "line of flight," movement that causes a temporary break in a structure. For Deleuze and Guattari this nomadic behavior, either literal or figurative, strips the assemblage of base and superstructure by flattening relations.[24] Deterritorialization is a positive for Deleuze and Guattari; to flatten is to exist horizontally, rhizomatically. Mobility is becoming. We should not dig down and follow roots in an effort to find answers, but instead chase a line of flight and see what happens. This has bigger implications for ideology, which, they bluntly claim, is a concept that does not exist.[25]

The rejection of ideology is problematic. Deleuze and Guattari are working toward a completely material philosophy here, but saying ideology does not exist does not make it so. As Žižek argues, distancing one's self from ideology does not prevent one from acting in accordance with ideology.[26] Deleuze and Guattari state that the answer lies in the "intermingling of bodies in a society," which ultimately is rooted in food and sex. Further:

> Even technology makes the mistake of considering tools in isolation: tools exist only in relation to the interminglings they make possible. . . . The stirrup entails a new man-horse symbiosis that at the same time entails new weapons and new instruments. Tools are inseparable from symbioses . . . defining a Nature-Society machinic assemblage . . . a society is defined by its amalgamations, not its tools.[27]

This strikes me as being far from coincidentally connected to Dant's driver-car assemblage. More importantly, however, is the fact that assemblages, symbioses, and amalgamations of bodies and objects do not preclude ideology. Bodies may assemble with machines to ultimately gain sustenance and reproduce, but such an assemblage does not produce drones driven by biological urges. Ideology, disconnected from reality as it maybe, is a factor in how bodies intermingle. And, most importantly, why must a line of flight, something Deleuze and Guattari hold as preceding urban limits and national borders, be free of ideology? *Fury Road* shows that mobility and ideology can be one and the same.

It is important to state that, despite all of this assemblage talk, hybridization also occurs in the film. While the assemblage is necessary for action and mobility, the hybridization of human and machine is also imperative for survival. Dant uses his assemblage concept precisely to reject the idea of a cyborg or hybrid relationship between humans and machine, but a film like *Fury Road* challenges the *temporary* nature of humans and machines working

Figure 2.1 Meeting Furiosa through Ideological Framing. *Source*: *Mad Max: Fury Road*, dir. George Miller (2015)

together. After Max has been captured and subdued by the War Boys, the prologue is over, and the film cuts to a close up of Imperator Furiosa's (Charlize Theron) neck as she walks away from the camera. She has been branded with the image of a skull-emblazoned, flaming steering wheel, therefore literally connecting her to automobility. We will learn that this is the symbol of Immortan Joe's War Boys, one tribe of many in this machine-driven world. The camera follows Furiosa, but slowly so that she pulls away, revealing her robotic prosthetic arm. The mise-en-scene is a blur of flesh and metal. We see that she is a cyborg but simultaneously her walking away reveals that she is moving toward several assemblages of the War Boys and their machines. To her left is one on a motorcycle; to her right is a devilish hot rod with a machine gun mounted to the top, the gunner standing at the back of the car. Another motorcycle and rider are just behind the car. Her "War Rig," a heavily armored semi-truck with human skulls adorning the front bumper, sits in the background (see figure 2.1).

The prologue establishes Max as needing his machine for survival, but this scene introducing Furiosa takes the ideology further buy blurring the lines separating human and machine. Yes, Dant's driver-car assemblage is relevant, but it is not the only thing at work in this scene. Furiosa is able to assemble with her truck because she has fused with a machine.

The scene cuts to a tanker trailer on a platform being lowered by a series of pulleys and massive chains. Pale War Boys, called "half-life" to denote their bodies decaying from the effects of the nuclear war, surround the trailer. These simple machines are another assemblage; human bodies increase their strength by assembling with the pulleys. The scene cuts to Furiosa climbing into the cab of her truck. She uses both her organic and her metal hand to place a steering wheel onto the steering column. The steering wheel matches the brand on her neck. As soon as the wheel is locked in, the scene cuts to a close up of Immortan Joe's hideously scarred and blistered back. A young half-life blows dust onto Joe's back while another holds his arm. Immortan Joe releases a deep, phlegmy cough. Cut to Furiosa's War Rig, the trailer is fully lowered and connected to her truck. Cut to Immortan Joe's back; a protective plastic plate is placed onto it. Cut to the trailer. Cut to Joe's front, his head is still cut out of the shot. The front plate is hooked to the back, his entire torso is covered. The front of his plastic vest is molded in the shape of a muscular torso, though his body is far removed from the idealized plastic abdominal muscles. His half-life assistants place an impressive codpiece at the bottom of the plastic. It is a large, round metal gear. Flames are welded to the back and a metal skull is in the center front. Cut to a long shot of the truck. War Boys are finalizing their work. Cut back to Immortan Joe's face as he places a breathing mask up to his mouth and nose. The mask is a mix of skeletal teeth and tubes, yet another hybrid of the organic and metallic, bringing oxygen into Joe's system. We will never see Immortan Joe without this mask and it will only be removed at the end of the film in a rather violent manner.

Miller is playing with Eisenstein's dialectical montage here. The ideology is not apparent in the individual frames but the collision of the machine being assembled by men and a man being turned into a machine produces something new intellectually.[28] Miller has blended human and machine so well that even the distinction between hybrid and assemblage becomes messy. Imaging these humans living free of machines becomes impossible. Dant's temporary driver-car assemblage explains the fusion of Max and his V-8 Interceptor. Deleuze and Guattari's fixation on horizontal movement explains the constant motion in which we find the film's characters. Most significantly though, is an underlying, permanent fusion of flesh and metal. Even if a character like Furiosa can physically step away from her War Rig, she still needs it to survive in her world.

Throughout the film this play of assemblage and hybrid, human and machine continues. Lactating women are hooked up to massive breast pumps to industrialize the production of milk. The platform that raises and lowers vehicles is comprised of machines, but it is driven not by burning hydrocarbons, but rather by slaves moving massive paddlewheels with their bodies. Max's "full-life" body will be strapped to a car to serve as a blood bag for the Half-life War Boy Nux (Nicholas Hoult). Nux has a picture of a V-8 car engine carved into his chest. Even the film's musical score is a hybrid of diegetic and non-diegetic sound. The hammering of percussion and guitar (electric of course) as Immortan Joe and his Half-life War Boys chase Furiosa pervasively fills the film's scenes, but is clearly emanating from one of the vehicles. The pounding of the drums matches the pounding of cylinders in the engines. Are these drummers in war or a film score? Is there a difference? Everything blurs. Again, the fact that Miller has made this film at the start of the Anthropocene is relevant. "Climate change makes the global atmosphere, its chemistry and weather systems, into Frankenstein's monster—part natural, part made."[29] Even nature is a hybrid.

SHIFTING INTO 4L: SPORT UTILITY VEHICLES, DISTANCING, AND CAPITALISM

The vehicles in *Fury Road* are not simply automobiles. They may have been prior to the oil and water wars, but not a single vehicle is stock. All of them have been modified to maximize power and weaponize the vehicles to maximize survival. Roads exist in the film's landscape, though none are paved. They are long stretches of compacted red sediment that demand off-road capabilities. The world of *Mad Max* is full of Sport Utility Vehicles (SUVs).

SUVs are seen as one of the fundamental problems with automobility. Mimi Sheller argues that these vehicles are prized more for their emotional rather than off-road capabilities. Even the SUV's habitat, nature, is produced:

> So-called "Sport Utility Vehicles" also continue to be embraced as a way of getting closer to nature (safely). Ironically, the very idea of "nature" that many anti-car campaigners are defending may have been constituted largely through automobility.[30]

Sheller argues that nature and industrial society are not as inseparable as it may seem. Of course, what Sheller does not include is that nature has always been produced by the cultures that inhabit it. European nature is a different space than American Indian nature, for example.[31] Nature is a hybrid.

Off-road automobility has also been given animalistic qualities. As Nicole Shukin writes, "The [Saturn] *Vue*—'at home in almost any environment'—is just one SUV among many eager to neutralize political antagonisms of automobility culture."[32] Giving an SUV an "environment" rather than placing it into the spaces of capitalism and transport cloud its materiality.

> An unabashed identification of automobile and animal emerges with the *Vue* ads. By equating automobility with the biological ignition of animal life, the *Vue* discourse mythologizes the motive power of the sport utility vehicle and conceals the economy of power regulating a carnivorously capitalistic relation of nature and culture.[33]

The mythologizing of which Shukin speaks is not unexpected in the ideology of automobility, nor ecomobility. If our mobility choices revolve around us travelling through "natural" space, does it not follow that humans would connect this mobility to more ecological terms? Further, the hybridized and social production of nature makes an ontologically pure wilderness impossible. It must be remembered that the ideologies of auto- and ecomobility should not be simply criticized for their obfuscation of the material world. Rather, one must ask, what work does such obfuscation do for automobility's subjects? Does not the blurring of boundaries resemble Deleuze and Guattari's deterritorialization? The driver of an SUV is not connected to the roots of modern environmentalism and the production of knowledge about nature, but rather moves in between urban and the wild spaces as a means of becoming. Nature, car, human assemble across horizontal space. This is not so much a critique of the concept of nature in the ideology of automobility (though this should continue to be questioned), but rather a critique of why the subject desires to drive through ecologies on the planet.

Mark Dery laments the state of American roads: "the bullying SUV and its even nastier, more brutish successor, the Hummer—two giant steps backward for fuel efficiency, passenger safety, and inconspicuous consumption—are the undisputed Kings of the Road in America."[34] Bullying, nasty, brutish—the SUV is a machine of violence, which Dery sees fusing psychologically with the driver.

The concern with the violence of SUVs and the nature in which they belong is a distraction though. Nature is clearly complicated, but so too violence is more complicated than the aggressive styling of an SUV. Returning to Žižek, we can split violence into three types: subjective, symbolic, and systemic.[35] Subjective violence is the overt violence of a film like *Fury Road.* When the dying half-life sprays silver paint into his mouth and demands his brothers witness his *kamikaze* destruction of the spikey "Buzzard" rig, we are taken by the spectacle of violence. While this violence is meant to entertain, it also distracts from what is additionally at work in the world of *Fury Road.* Žižek's systemic violence is the most interesting form in connection with automobility. This is the violence that occurs when structures and systems *are running as they should.* A new breed of capitalist, what Žižek calls a "liberal communist" works to stop the subjective violence of poverty, disease, malnutrition, all the while masking that the ideology of capitalism is responsible for those very problems. "The same philanthropists who give millions for AIDS or education in tolerance have ruined the lives of thousands through financial speculation and thus created the conditions for the rise of the very intolerance that is being fought."[36] It is not that Žižek is discounting the severity of the AIDS epidemic; this is a horrible disease that should be eradicated. The problem is that the capitalists giving to the research to fight the disease are using the ideology of humanitarianism to mask the violence of the capitalist ideology.

How does this connect to automobility? Whether we are discussing the subjective violence of aggressive SUV styling in contemporary America or the subjective violence of *Fury Road's* characters on their motorcycles, monster trucks, and the War Rig, we run the risk of ignoring or overlooking the systemic violence of automobility. For modern American automobile choices, the critique of the bullying SUV masks the subjective violence of the Prius. Again, fuel efficiency does not reduce fuel consumption, but the Lacanian Real of the ideology of automobility tells us that by purchasing more hybrid vehicles we are saving the planet. Per Žižek, "'reality' is the social reality of the actual people involved in interaction and in the productive processes, while the Real is the inexorable 'abstract,' spectral logic of capital that determines what goes on in social reality."[37] Ideology is not something a subject uses to escape reality, but instead comes from the Real to allow a subject to make sense of reality. Further, ideology does not fully succeed in grabbing the subject in an Althusserian sense. We need not fully embrace automobility for it to be durable. Purchasing a hybrid automobile would appear to be a failure of a perfect submission to the ideology, yet Žižek sees that gap between interpellation and subject as the very site of subject formation.[38] Ideological fantasy works to structure reality. We know very well that occasionally using electricity to move a vehicle does not erase its impact on the environment, but automobility tells us that there is no other reasonable way in which to move across that environment. It is not that reality fails to provide alternatives, but rather, the Real of automobility "succeeds in determining the mode of our everyday experience of reality itself."[39] In *Fury Road* the subjective violence that seemingly drives this action film, actually gets in the way of a true critique of the Real of automobility. There are hints at the ideological violence that caused the world of *Fury Road* throughout the film. The oil and water wars, the nuclear destruction, the roving tribes of brutish half-lives driving equally brutish off-road vehicles are simply the most visible forms of violence. Underlying all of this is capitalism.

Early in the film, Immortan Joe reaches his vault and sees that his prized breeding wives are gone. Graffiti, we are to assume were written by the wives, covers the walls. "WE ARE NOT THINGS" says one. "OUR BABIES WILL NOT BE WARLORDS" says another. But most importantly "WHO KILLED THE WORLD?" is scrawled over the vault's entrance. This same question is asked to Nux by Immortan Joe's favorite wife, Splendid (Rosie Huntington-Whiteley), later in the film. It would be easy to read this as a feminist environmental message of the Gaia variety. Men killed the world by poisoning the atmosphere. But, when placed alongside the other images of the film, it becomes more likely that it wasn't men *qua* men that killed the world, but rather men *qua* capitalists that did so. The commodification of resources—water, oil, wombs—is the real villain of the film, despite the clear subjective violence of Immortan Joe.

Capitalism and specifically the formation of commodities drives automobility. In the scene in which Max first walks up to Furiosa and the wives, he demands they bring him water. Splendid, who is clearly pregnant, carries a dripping hose. The shot is framed so that we see her swollen belly and the hose together. Both the child and the water are commodities to be used by men. Immortan Joe holds power because his home sits atop an aquifer. Water is not even called water by Joe, but instead "Aqua-Cola" to fully make it a commodity and something transcendent.[40] "Because he owns it, he owns all of us," explains one of his wives. His babies have a use value in that they can grow up to fight other tribes. All of this revolves around the vehicles that must be driven to Gas Town for more "guzzoline" (gasoline) and the Bullet Farm for more ammunition. One can no longer travel the Earth without the aid of a powerful and aggressive automobile. This Earth demands automobility of a specific kind, which means commodities drive the world of *Fury Road.*

One of the more disturbing characters in the film is The People Eater, a grotesque man who must be lifted into his vehicle by a number of half-lives. He is an accountant ensuring that capitalism is not forgotten. As Immortan Joe pursues his escaped wives, The People Eater reminds him, "We are down 30,000 units of guzzoline, 19 canisters of nitro, 12 assault bikes, seven pursuit vehicles. The deficit mounts, and now sir, you have us stuck in a quagmire!" The subjective violence is fine as long as it does not impact the Real of capitalism. The gasoline burning off-road vehicles are necessary for survival; but the violence of capitalism killed the world.

Another of Joe's escaped wives, The Dag (Abbey Lee), speaks to The Keeper of Seeds (Melissa Jaffer), one of the Many Mothers of The Green Place. The Keeper of Seeds makes a comment about "snapping" people, meaning shooting them. The Dag responds to this display of yet more subjective violence by saying that these women should be above it. This is how we know that the film isn't just a feminist message. The Keeper of Seeds shows The Dag her bag of seeds and explains what they are. "Trees. Flowers. Fruit. Back then, everyone had their fill. Back then, there was no need to snap anybody." When resources were plentiful, subjective violence did not have to exist. When resources grew scarce, and men like Immortan Joe hoarded them, subjective violence became necessary. The violence of capitalism killed the world.

THROTTLE DOWN: CONCLUSIONS

Mad Max: Fury Road is a violent and aggressive film, but that should not distract from the systemic violence of capitalism and automobility. The ideology of automobility, specifically excessive American automobility, may appear impossible to sustain, but it avoids crises through the ideological fantasy that masks the violence it commits upon the Earth and upon bodies. The absurdity of *Fury Road*, the over-the-top vehicles and characters assembling for mobility, lay bare the ideology of both automobility and ecomobility. While Miller's overt violence committed by his characters distracts from the systemic violence, he nonetheless has made a film that uses absurdity to force the viewer's awareness of ideology. Ecomobility is revealed in the need for humans and machines to assemble to travel over the specific ecosystems of the film. Figure 2.2 shows Max positioned between the War Rig that got him to what was left of The Green Place and the motorcycle that could take him toward his next destination. The monstrous styling of the War Rig and the ruggedness of the motorcycle strips the subtlety of Miller's message. The red sand cannot be traversed without one of these machines. An assemblage must occur for Max to survive as an organism within this ecosystem. The burning of fuel is his ecological niche. The underlying ideological need for machinic mobility is not questioned.

Automobility's ideology is also presented through the images and dialogue in *Fury Road*. Humans and machines assemble without questioning the need to do so. This is more than assemblage in Dant's use of the word, in which drivers and cars temporarily fuse as needed. Humans and machines also do not assemble to push for deterritorialization as Deleuze and Guattari would

Figure 2.2 The Ecomobility Assemblages of *Mad Max: Fury Road*. *Source*: *Mad Max: Fury Road*, dir. George Miller (2015)

have it. Humans and machines and their environments are hybrids that at least appear to require a symbiotic relationship. Humans need machines to exist in the environment of *Fury Road.* Yet that apparent need for automobility gets at the ideological fantasy at work. The outlandish nature of this assemblage and hybridization reveals just how far we distance ourselves from modernity's regime of automobility. We see how over-the-top Immortan Joe's fleet of vehicles are, which at first glance might reveal the problems of automobility. Yet, The People Eater shows that capitalism commits the real violence in the assemblage. Automobility exists for profit. The question we are left with is what does a resistance of automobility look like? Certainly not a Prius. Resistance does not mean an end to internal combustion, but instead a questioning of commodification. The women returning to and taking over Immortan Joe's Citadel are not ending the assemblage/hybridization of human and machine. We are to assume that water will now be shared. The end of the commodification of natural resources is what this violent road trip accomplished. Max of course disappears into the desert; his mobility cannot be arrested.

NOTES

1. K. Smith, "Why 'Mad Max: Fury Road' is the Feminist Picture of the Year," *New York Post*, 14 May, 2015. http://nypost.com/2015/05/14/why-mad-max-fury-road-is-the-feminist-picture-of-the-year/; P. Chin, "The Problem with 'Mad Max' Feminism," *Christianity Today*, 3 June, 2015, http://www.christianitytoday.com/ct/2015/june-web-only/problem-with-mad-max-feminism.html; A. de Coning, "Recouping Masculinity: Men's Rights Activists' Responses to Mad Max: Fury Road," *Feminist Media Studies* 16, no. 1 (2016): 174–176; Y. Klangwisan, "Mad Max Fury Road: Escaping the Phallic Economy of the Exodus," *The Bible and Critical Theory* 12 (2016): 75–86; A. Lane, "High Gear: 'Mad Max: Fury Road'," *The New Yorker*, 25 May, 2015, http://www.newyorker.com/magazine/2015/05/25/high-gear-current-cinema-anthony-lane; S. Hawk, "The Fury Behind Fury Road," *Sierra*, 9 April, 2015, http://sierraclub.org/sierra/2015-3-may-june/feature/fury-road-all-your-darkest-environmental-nightmares-come-true.
2. John Urry, "Inhabiting the Car," *The Sociological Review* 54, no. s1 (2006): 18.
3. Urry, "Inhabiting," 18.
4. S. Böhm, C. Jones, C. Land, and M. Paterson, "Introduction: Impossibilities of Automobility," *The Sociological Review* 54, no. s1 (2006): 3.
5. Michel Foucault, *The History of Sexuality, Volume 1: An Introduction* (New York: Vintage, 1990 [1978]), 95.
6. Böhm et al., "Introduction," 8.
7. Paul Groth, "Frameworks for Cultural Landscape Study," in *Understanding Ordinary Landscapes*, edited by P. Groth and T.W. Bressi (New Haven: Yale University Press, 1997), 1.

8. E. Glaeser, *Triumph of the City: How Our Greatest Invention Makes Us Richer, Smarter, Greener, Healthier, and Happier* (New York: Penguin Books, 2011); J. Speck, *Walkable City: How Downtown Can Save America One Step at a Time* (New York: Farrar, Straus, and Giroux, 2012).

9. Slavoj Žižek, *The Sublime Object of Ideology* (London: Verso, 2008 [1989]), 25.

10. Sergei Eisenstein, *Film Form: Essays in Film Theory* (New York: Harcourt, Brace, and World, Inc., 1949), 49–52.

11. J.R. McNeill and P. Engelke, *The Great Acceleration: An Environmental History of the Anthropocene since 1945* (Cambridge: The Belknap Press of Harvard University Press, 2014), 2.

12. Böhm et al., "Introduction," 9–11; Urry, "Inhabiting," 18.

13. D. Owens, *The Conundrum: How Scientific Innovation, Increased Efficiency, and Good Intentions Can Make Our Energy and Climate Problems Worse* (New York: Riverhead Books, 2011), 95; J. Linn, *The Rebound Effect for Passenger Vehicles* (Washington, DC: Resources for the Future, 2013); Also see R. Dowling and C. Simpson, "'Shift-the Way You Move': Reconstituting Automobility," *Continuum* 27, no. 3 (2013): 427.

14. Urry, "Inhabiting," 18.

15. Žižek, *Sublime Object*, 30.

16. Paul Kingsbury, "Uneasiness in Culture, or Negotiating the Sublime Distances Towards the Big Other," *Geography Compass* 11, no. 6 (2017): 8.

17. Slavoj Žižek, *The Plague of Fantasies* (London: Verso, 2008 [1997]), 27.

18. Böhm et al., 9.

19. Owens, *Conundrum*, 102.

20. Tim Dant, "The Driver-Car," *Theory, Culture & Society* 21, no. 4/5 (2004): 62.

21. Dant, "Driver-Car," 75; Gilles Deleuze and Felix Guattari, *A Thousand Plateaus: Capitalism and Schizophrenia*, translated by Brian Massumi (Minneapolis: University of Minnesota Press, 1987 [1980]), 88.

22. Deleuze and Guattari, *Plateaus*, 89.

23. Deleuze and Guattari, *Plateaus*, 9 and 88–89.

24. Deleuze and Guattari, *Plateaus*, 90.

25. Deleuze and Guattari, *Plateaus*, 4.

26. Žižek, *Sublime Object*, 30.

27. Deleuze and Guattari, *Plateaus*, 90.

28. Eisenstein, *Film Form*, 82–83.

29. J. Purdy, *After Nature: A Politics for the Anthropocene* (Cambridge: Harvard University Press, 2015), 15.

30. Mimi Sheller, "Automotive Emotions: Feeling the Car," *Theory Culture Society* 21 (2004): 231.

31. C. Coté, *Spirits of our Whaling Ancestors: Revitalizing Makah and Nuu-chah-nulth Traditions* (Seattle: University of Washington Press, 2010).

32. Nicole Shukin, "The Mimetics of Mobile Capital," *The Sociological Review* 54, no. s1 (2006): 159.

33. Shukin, "Mimetics," 159.

34. Mark Dery, "'Always Crashing in the Same Car': A Head-on Collision with the Technosphere," *The Sociological Review* 54, no. s1 (2006): 223.
35. Slavoj Žižek, *Violence* (New York: Picador, 2008), 1–2.
36. Žižek, *Violence*, 37.
37. Žižek, *Violence*, 13.
38. Žižek, *Sublime Object*, 43.
39. Žižek, *Sublime Object*, 49.
40. Karl Marx, *Capital: A Critique of Political Economy, Volume One* (London: Penguin Books, 1990 [1867]), 163–164.

Chapter 3

Machines Precede the Climate

The Technological Fix

THE CLIMATE IS HYPERREAL

Scholars of the Anthropocene have introduced the hybridity of the Earth and a fusion of nature/society that resists past divisions.[1] As the previous chapter discussed, the hybridization and assemblages of the driver-car offer new perspectives of what such a blending of nature and humanity might look like.[2] But, just as the world of Mad Max struggles to envision mobility without big engines and fossil fuels, I want to use this chapter to explore stories that reveal, either wittingly or not, the problems with global scale government and private enterprise technology solutions. Turning to Marx, the unwillingness to see the labor inherent in our machines that we assemble or fuse with suggests that waiting for a technological fix to climate change is absurd.

David Harvey points to a small footnote in the first volume of *Capital* that he argues is Marx providing a subtly complex framework for further study.[3] Harvey breaks the note into three distinct parts, with the second being:

> Technology reveals the active relation of [hu]man to nature, the direct process of the production of [their] life, and thereby it also lays bare the process of the production of the social relations of [their] life, and of the mental conceptions that flow from those relations.[4]

Harvey pulls out a wealth of information from this one sentence, arguing for an entire six-part assemblage of technology, nature, modes of production, reproduction of daily life, social relations, and mental conceptions of the world.[5] Marx's use of "reveals" is not causal, but implies a dialectal relationship across this matrix.

> What Marx is saying (and plenty of people will disagree with me on this) is that technologies and organizational forms *internalize* a certain relation to nature as well as to mental conceptions and social relations, daily life and labor processes. By virtue of this internalization, the study of technologies and organizational forms is bound to "reveal" or "disclose" a great deal about the other elements.[6]

Harvey pushes back against accusations that Marx was a technological determinist and argues instead that studying technologies in a specific society at a specific moment in time will offer insight into the rest of society. Harvey directly links his read of Marx here to a Deleuzian assemblage, though he still insists on the term dialectic.[7] His point is the open relationship of the elements, rather than having one determine the rest, but I think more emphasis on a Deleuzian "and . . . and . . . and . . ." is important in fully realizing this framework.[8] We are not so much opposing these facets of humanity as we are setting them atop one another like geologic strata to build something new. It is with this framework in mind that I want to focus on how our machines and mobilities reveal both a social conception of and the material consumption of nature, as well as the reproduction of capitalist relations. Rather than a purely Marxist read, however, I want to bring Baudrillard's simulacrum into the assemblage to trouble the connection between technology, humans, and nature. The machines precede the climate.

As Baudrillard contends, "It is nevertheless the map that precedes the territory."[9] He is speaking of Borges' rather short story of the map of an empire being drawn at a scale of 1:1.[10] The map covers the entirety of the empire and as it collapses, so too does the map. Baudrillard argues that such a story is no longer relevant in a hyperreal world, that is, one that is not comprised of representations of reality. The very abstraction needed to move between the real and the represented is missing today. "Simulation" is Baudrillard's term to replace representation; "whereas representation attempts to absorb simulation by interpreting it as a false representation, simulation envelops the whole edifice of representation itself as a simulacrum."[11] Simulation is the act of pretending that something exists when it does not, while the simulacrum exists after the simulation when there no longer is a connection between the image and a real object. So the term hyperreal describes how simulacra are not tied to a "true" reality and are therefore their own type of real. With the Borges map example, the map now "precedes" the territory and is real all on its own.

Baudrillard is often mocked for such an insistence that reality is not real, I think by people who cannot help but bring a positivist approach to such statements (it probably has to do with what I find to be an incredibly challenging writing style as well). I invoke the man because a healthy questioning of reality and representation is needed at a moment like this. Between the climate crisis, COVID-19, and a Donald Trump presidency, exactly what is reality anyway? And while this question might seem absurd, we must stop and ask does the term reality have any value in how Americans have tackled climate change.

I want to use this chapter to tarry with Baudrillard's relationship between reality, simulation, and simulacra in the context of machines, mobilities, and Earth. Specifically, I think it is rather important to examine films like *Pacific Rim*, *I Am Mother*, and *Snowpiercer* to get at Baudrillard's point and get a sense of how we are stuck in a positive feedback loop of using technology to fix technology's side effects. Films are the "very definition of the hyperreal."[12] My thought is that machines precede the climate. I will hold off in stating the climate crisis did not take place,[13] mainly because that would be the one sentence taken from this while missing my point, but it is important to question if our response to climate change is actually connected to "the real." The rush to invent new machines to replace the ones that got us into this mess is a short circuit that bypasses climate altogether to maintain economic structures. This is yet another invocation of Mark Fisher's *Capitalist Realism*, in that due to our lack of imagination we cannot escape capitalism and thus cannot imagine fixing climate change without working within capitalism.[14] My fear is that by working within capitalist assumptions about the progress of machines, we have eradicated any connection to living in nature. I want to follow Baudrillard's claim that "it is *now impossible to isolate the process of the real*, or to prove the real" and see how the self-referential world of cinema has affected our knowledge about dealing with the climate crisis.[15]

PUNCHING AND KICKING CLIMATE CHANGE: *PACIFIC RIM*

Guillermo del Toro's *Pacific Rim* is chock full of machines. Big humanoid machines that punch and kick big monsters. Both the monsters and the machines have a bright Christmas Tree palette of reds, greens, ambers, and blue lights. It's a colorful, loud, ridiculous movie that was generally well received and spawned a sequel.

The film opens with a black screen and two definitions written in a green computer style font: *Kaiju* a Japanese word meaning giant beast, and *Jaeger*, a German word for hunter. The first word refers to the alien life that has erupted from a tectonic portal at the bottom of the Pacific Ocean, giant monsters that move toward coastlines and destroy civilization. The Kaiju have no allegiance of course, and attack cities in nations on every edge of the Pacific. To fight the monsters, our narrator explains, we humans "had to create monsters of our own" by building giant robots designed to essentially just punch and kick the Kaiju (Figure 3.1). Have I mentioned that the film is ridiculous?

This blending of the Japanese and German languages hints at the overall theme of the film which is the merging of seemingly different groups to accomplish something for the good of humanity. Del Toro himself has said as much:

> "The pilots' smaller stories actually make a bigger point, which is that we're all together in the same robot [in life]," he says, chuckling a bit at the sound of this. "Either we get along or we die. I didn't want this to be a recruitment ad or anything jingoistic. The idea of the movie is just for us to trust each other, to cross over barriers of color, sex, beliefs, whatever, and just stick together. Fundamentally, it's a very simple message. But it's one that I would have liked to have seen in an adventure movie when I was a 12-year-old."[16]

The Jaegers need to be piloted by two humans who "drift," a process of blending both minds into one to be able to control such a sophisticated machine. What is fascinating about this assemblage is that it takes something like Dant's driver-car to the next level.[17] The pilot (driver) must link their brain with another human who is an acceptable match so that both can fuse with the machine. The mentally conjoined pilots do not simply turn a steering wheel but rather move as they normally would which then tells the machine what to do. As Dant says, "the driver-car is neither a thing nor a person; it is an assembled social being that takes on properties of both and cannot exist without both."[18] These pilots not only move through the world in a way in which they could not with the aid of the Jaeger–the humanoid structure of the Jaeger produces an object that extends human capabilities in a way the automobile cannot. Where a critic like Dant rightly points to the difficulties of abandoning the automobile as we know it because of our social, physical, and emotional connection to the machines, *Pacific Rim* doubles down and shows that our troubles come from not fusing with the machines enough.

The Kaiju are a perfect allegory for the climate crisis in that they are a destructive force seemingly produced of the Earth that threatens all of humanity while not being as abstract a concept like gradual temperature rise. The fact that the monsters are just that, monsters, that can be punched and kicked by giant machines make this a story of humans triumphing over nature through technological progress. Early in the film, the protagonist Raleigh Becket (Charlie Hunnam) explains that "there are things you can't fight, acts of god. You see a hurricane coming, you have to get out of the way. But when you're in a Jaeger, suddenly, you can fight the hurricane." That line might be the best single statement to describe an American approach to living in nature.

This active approach of using Jaegers to battle that which comes out of the Earth is under another threat, one of governmental oversight and budget limitations. An international panel of world leaders has decided to cancel the Jaeger program to replace it with a passive "Coastal Wall" protecting the different cities by simply erecting a tall barrier.[19] Moments later we see news footage of a Kaiju effortlessly breaking through the wall protecting Sydney, Australia. Fortunately, a Jaeger is there to defeat the beast. Clearly, humanity needs advanced mobile machines, not ancient static structures to ward off such monstrous threats.

So far, none of what I have described is a brave new read of *Pacific Rim*. Even the most disinterested viewer can pick up these themes. At one point in the film it is explained that the beings responsible for sending the Kaiju to Earth's surface are colonizers and that our own pollution of our planet made it more hospitable for them. Clearly, environmental degradation is embedded in the film. There are two underlying components of the film, however, that I think are crucial in understanding our continued failures to punch and kick the climate crisis. The first is the surprising lack of new energy sources to both power the Jaegers and to destroy the monsters. The machines are powered by diesel engines or nuclear power and the big plan is to release a nuclear bomb at the portal at the bottom of the ocean to once and for all stop the Kaiju from wreaking havoc on Earth. In fact, the more advanced digital Jaegers are rendered useless by a new Kaiju weapon and it is up to the older analog model to save the day. In the absence of some magic bullet type energy source like we see in other science fiction movies, the secret to humanity's success is not getting hung up on nationalism, sex, or any other identity and instead coming together. While the sentiment is well intentioned, the very absence of new ideas shows that we humans are doomed. If we read the film as instructions on dealing with an alien menace, fine, but if we instead read the film as a climate allegory, *Pacific Rim* shows that we have yet to see that our hunger for energy and even bigger machines is the actual threat. As the Jaegers effortlessly move over and through natural spaces, one can easily picture an SUV or truck commercial of a rig conquering a barren mountain.

The second subtle but damning element of *Pacific Rim* is also the most entertaining aspect. As soon as Kaiju appeared, so too did a black market revolving around the harvesting and selling off of dead Kaiju parts. Hannibal Chau (Ron Perlman) runs the black market for all of Asia, which we learn is in collaboration with the Jaeger program. There is no mystery in his role in the film. Chau mocks those who believe the Kaiju are a punishment sent by God. When asked what he believes, Chau responds, "Well, I believe that Kaiju bone powder is $500 a pound." He clearly is a capitalist making the most of the Kaiju crisis who does not give much thought about what happens later. Hannibal Chau reveals the capitalist realism inherent in *Pacific Rim*.

Figure 3.1 Jaegers Effortlessly Move through Nature on Their Way to Battle Kaiju in *Pacific Rim*. *Source*: *Pacific Rim*, dir. Guillermo del Toro (2013)

The film has no problem envisioning the end of the world, but capitalism is thriving in this crisis. Chau's black market operation of harvesting and selling off Kaiju parts is the epitome of moving capital to seize new global markets. Hannibal Chau does get eaten by baby Kaiju in the film, yes, but those who stuck through the film's credits saw an extra scene in which the deceased baby Kaiju is split open by Chau's trademark butterfly knife. He crawls out of the slimy slit and asks, "Where's my goddamn shoe?" referencing one of his gold tipped oxfords that flew off when he was eaten. The credits resume. While we can eradicate Kaiju, capitalism cannot die.

STARTING FRESH: *I AM MOTHER*

The film's premise of human extinction is perhaps the most extreme of the bunch, in that Grant Sputore's *I Am Mother* begins once the extinction has occurred. It opens in a dimly lit, futuristic room with the words "UNU-HWK Repopulation Facility" across the screen. The titles explain that the extinction event has just happened and that 63,000 human embryos are housed within this facility. A humanoid robot, one that has a rather early 2000s Apple vibe, is assembled by robotic arms. Once up and running, the robot, Mother (performed by Luke Hawker and voiced by Rose Byrne), goes to work selecting the first embryo to grow to begin the task of repopulating the Earth. Despite the mechanical appearance of Mother, *she* (what a great example of using a pronoun to connote performative gender rather than biological sex) exudes a warm, motherly care as she pulls the fully gestated baby from a floating artificial womb. As the baby sleeps, she wraps her small finger around one of Mother's robotic digits. Warming lights on the robot's chest and arms show the audience that Mother was built to care for human life. A montage of scenes with the baby growing into a toddler continues to show just how loving Mother is, even if the idea of robots and love seems incompatible. And it works, the opening of the film is incredibly sweet despite all of the cold metal and plastic. The human-like qualities of the machine convince us that she does love this child, and as we watch the baby grow into a young girl, it is clear that the girl loves this machine like a human mother. When the girl asks Mother why she only grew one out of the thousands of embryos, the robot explains that "Mother's need time to learn. Raising a good child is no small task." This feminine assemblage with machines is fundamentally different from automobility and yet, once again reveals a naturalization of technology. In the absence of a human mother, of course we will simply build a robotic one out of plastic and metal. The initial scenes of the baby growing into girlhood play the cold dark metal spaces of this repopulation facility against the warmth and literal glowing of Mother.

The film cuts to flashing computer displays and the title "Days since the extinction event: 13,867)" followed by "Current Human Occupants: 001." The camera pedestals down to show the sleeping face of a teenage version of that young girl, referred to as Daughter (Clara Rugaard), we had just been watching. We are to assume time has passed and she has grown. Doing the math, however, reveals that the extinction event took place almost 40 years ago. Daughter is not the same girl we just watched grow up. However, thanks to *montage*, those in the audience not taking the time to divide days into years will not have any doubt that we are still watching the same child. And the love shared between Mother and Daughter does not suggest that we should be suspicious. In one scene, Daughter sees that Mother's hand is not functioning properly and repairs it for her. Again, we see that the human machine assemblage works both ways in a mutualistic relationship. Both entities need the other to function. They are clearly symbiotic.

Daughter receives an education from Mother. We are to assume it is a well-rounded one, though we only witness a lecture on utilitarian ideals and the thought of Jeremy Bentham. Bentham's principle of utility, which argues that any action in society of both governments and individuals, ought to be weighed by "the tendency which it appears to have to augment or diminish the happiness of the party whose interest is in question: or, what is the same thing in other words, to promote or to oppose that happiness,"[20] is a guiding theme throughout the film. The notion of providing the least harm to the most people is explicitly discussed here, but is immediately put into action when Daughter catches a small rat crawling around the facility. Mother is concerned that it might carry the virus we are to assume led to the death of humanity so she callously throws it into an incinerator. Despite Daughter's obvious infatuation with the small mammal, Mother destroys it to insure that nothing risks her life, or the lives of the unborn fetuses in the facility. It is a lesson that Daughter takes to heart.

One day a wounded Woman (Hilary Swank) knocks on the facility door. Mother is on the other side of the facility, so Daughter is able to let Woman inside. Of course, Woman's presence shatters Daughter's conception of the outside world. Not only is there at least one other human in the world, but Woman explains that there is no contagion outside that threatens human life. Further, Woman gets worried when she realizes that there is a robot inside the facility. Daughter refers to her as Mother, but Woman uses the term "dozer" and the gender neutral "it." She also reveals that there are more robots outside the facility who were the things that wounded her.

Once Mother discovers Woman, she helps her in the infirmary after confiscating her shotgun. Mother, invoking Bentham's harm reduction, admits to Daughter that she has been lying about the virus outside, but that it was to keep her safe inside the facility. While Mother seems to still be a nurturing sentient being, Woman tells Daughter that the robots are the reason humans are gone and have committed atrocities to babies and families outside of the facility. Daughter is caught between the two stories. To make things more complicated, Mother allows Daughter to select a new embryo to be grown and born, this one a boy. Despite the joy surrounding a new brother, horror results when Daughter discovers that she herself is fetus APX03, the third to be born in the facility. APX01 and APX02 were grown but failed the ethical and psychological tests set for them by Mother and were therefore "aborted" and burned in the same incinerator where the rat met its fate. Even if the decision means leaving her new brother, Daughter feels she must escape with Woman, with the thought that they will get help from the other humans in the mines from where Woman came. Shortly after the escape we learn that Woman is lying and has not lived with any other humans for years. It seems Daughter cannot trust human nor machine.

When Daughter and Woman escape, we finally see the outside world. It is bleak. Dead, limbless trees erupt out of a smoky wasteland. The Sun is low in the horizon and barely breaks through the dirty atmosphere. Drones fly through the sky and other machines grow massive fields of corn. Woman explains to Daughter that the machines and the crops only showed up about six months ago; prior to their arrival it was almost impossible to breathe the air. Clearly, the Earth is being prepared for a return of humans. As Woman and Daughter move away from the terraforming, they pass downed powerlines and still jack pumps. Woman does not take them to the mines of which she spoke, but instead to the coast where a wrecked container ship has

Figure 3.2 The Remains of Capitalism in *I Am Mother*. *Source*: *I Am Mother*, dir. Grant Sputore (2019)

provided shelter (figure 3.2). In just a few shots, we realize that the only virus responsible for humanity's death was capitalism. It would seem that Director Grant Sputore has done it; we are gazing at the end of capitalism.

Shortly after arriving, Daughter leaves Woman and returns to the repopulation facility. Mother welcomes her back and explains that she is actually the artificial intelligence (AI) behind every robot on the planet. Her directive was to repopulate the Earth, but not simply through raising babies. Mother's task was to raise a smarter, more ethical human who could then go on to raise future generations and not make the same mistakes of the past. Of course, she had to eradicate humanity first in order to "elevate her creators." She still values human life above all else, just as she was originally programmed, but Mother has taken utilitarian ethics to heart: "More humans will flourish in the new world than ever perished in the old." Daughter takes her baby brother and "kills" the robot Mother, though her AI consciousness still exists in all machines on the Earth.

One of the other robots, speaking with Mother's voice, tracks down Woman back at the coast and explains that she was but a pawn in Mother's grand plan to perfect humanity. Since she is no longer needed, the robot presumably murders her. We are to assume that the only two humans on the planet are now Daughter and Brother. Further, Daughter appears to have no problem with Mother's ultimate plan and will rebuild a proper humanity.

I Am Mother is dystopian and at the same time shows that, through the "right" ethics, humanity can flourish. What is interesting here is that human flourishing takes the seemingly cold calculations of a robot to make the difficult choices to give humans a second chance. Where *Pacific Rim* shows a solution to our ability to live in nature is dependent on a more robust assemblage with our machines, *I Am Mother* argues that perhaps our only hope is to let the machines destroy us and then rebuild both humanity and a human/nature hybrid ecosystem.

While we have witnessed the end of capitalism in *I Am Mother*, it is disheartening that it took the end of humanity to get there. The alternative proposed by Mother, however, does not preclude the reemergence of a capitalist economy. Rather, Mother is creating a world in which any capitalist market would be driven by utilitarian ethics. Yes, some people will still suffer, but the fewest amount possible and for the greater good of the rest. What *I Am Mother* presents is the desperate need we humans have in returning to Enlightenment era principles. At its core, this film is a conservative dream of traditional ideas and an elite new world order. If we are to survive the climate crisis, we must look to the past when humans were more rational and therefore just beings. Capitalism isn't the problem, nor are machines. It is the fault of those irrational humans that just won't do the right thing.

"BABIES TASTE THE BEST": *SNOWPIERCER*

Bong Joon Ho's *Snowpiercer* is the darkest of the three films; perhaps due to the fact that it explicitly deals with class, a recurring theme in Bong's work. The film opens with audio clips of news reporters announcing the international geoengineering plan to disperse "CW-7" into the upper atmosphere to filter solar radiation and combat global warming. There is a quick shot of white contrails exiting airplanes overlaid with the ominous music growing louder. The film then cuts to a dark, frozen wasteland of still automobiles forever locked in traffic with

SOON AFTER DISPERSING CW-7
THE WORLD FROZE
ALL LIFE BECAME EXTINCT

written across the screen. A train rumbles past and the titles explain that the few survivors hopped aboard to escape the new ice age.

So while technology and scientific knowledge ruined the Earth, we will learn that the train, named *Snowpiercer*, is a technological marvel. It is fully contained and never stops moving. Water is recycled for survival but also fills elaborate saltwater aquarium tanks to provide fresh sushi. Verdant gardens produce fresh produce. The train has everything humanity could need within it; the frozen planet is but an inconvenience as long as these remaining humans keep moving. Time and space are also now linked in a new way.

The train takes one calendar year to circumnavigate the globe, so time is measured by the passing of key landmarks. Holidays and celebrations occur at key points along the track.

It is important to clarify that we learn all of this about the train as the film progresses. Unlike our current economic system, in which we are meant to see the wonders of capitalism and none of the horrors, *Snowpiercer* brings us into this world through the lives of those at the bottom that are brutalized by the comfortable façade.

After the explanation of the geoengineering disaster, the film leaps forward seventeen years and two heavily armed and armored men enter a train car to conduct a bed check. The camera follows one man and reveals a huddled mass of humanity, filthy and nervous. As the guard counts, each subsequent row of people sits down. Clearly the new climate has not brought the people of the world together as *Pacific Rim* suggests.

The train is segregated by class. The poor are relegated to the tail section of the train to live in crowded, dirty bunks while the wealthy enjoy a worry-free existence at the front. Armed guards and locked doors maintain the hierarchy. Because the film starts at the tail end of the train, the actions of the elites are as mysterious to us as they are to the poor passengers. Guards and officials dressed in nice clothing enter the car to select and remove passengers at whim. A man is taken to play the violin in the front of the train. Children are measured and then carted off to never be seen again. When the poor fight back they are violently punished, including having their bare limbs held out of the train to freeze in the subzero temperatures to then be broken off by a hammer. All of this is done in name of maintaining the proper order of things, explains Minister Mason (Tilda Swinton):

> Order is the barrier that holds back the frozen death. We must all of us, on this train of life remain in our allotted station. We must each of us, occupy our pre-ordained particular position. Would you wear a shoe on your head? Of course you wouldn't wear a shoe on your head. A shoe doesn't belong on your head. A shoe belongs on your foot. A hat belongs on your head. I am a hat, you are a shoe. I belong on the head, you belong on the foot. Yes? So it is.

Minister Mason explains that the initial tickets purchased by the train passengers are what initiated the hierarchy, but she then seamlessly naturalizes this order. "Eternal order is prescribed by the Sacred Engine . . . . All things in their place. All passengers in their section. All water flowing, all heat rising, pays homage to the Sacred Engine." This is pure Gramscian hegemony of common sense at work.[21] It might feel heavy-handed to the viewer, in part due to Tilda Swinton's excellently absurd performance, but isn't that how the best consent works? Is the emotional call connecting people remaining in their places to physical laws any different from the car commercials placing an SUV at home in nature? We can invoke Žižek here: I know perfectly well that an SUV isn't natural, but still . . . [how else could I get out into the wild]?[22] I know perfectly well that predetermined castes make no sense, but still . . . [some people just belong at the bottom of society].

As the film progresses, so too do the links to Gramsci. *Snowpiercer*'s plot follows Curtis Everett (Chris Evans) as he leads a revolutionary force of lower class passengers in their effort to overtake the front of the train. Curtis and his mentor, Gilliam (John Hurt), as leaders of the revolution, can be read as organic intellectuals, those people of the working class not fulfilling traditional societal roles, but instead tasked with bringing consent to a new revolutionary group. As Gramsci states:

> One of the most important characteristics of any group that is developing towards dominance is its struggle to assimilate and to conquer "ideologically" the traditional intellectuals, but this assimilation and conquest is made quicker and more efficacious the more the group in question succeeds in simultaneously elaborating its own organic intellectuals.[23]

Where Minister Mason works to convince those in the rear car to stay in their places, Curtis must work even harder to convince them to rise up, while not becoming the oppressor. Curtis' approach and challenges are apparent in his exchange with Edgar (Jamie Bell):

Edgar: Those bastards in the front section think they own us. Eating their steak dinners and listening to string quartets and that.
Curtis: We'll be different when we get there.
Edgar: [sighs] I want a steak.

Curtis has his sights on a true revolution, whereas many of his comrades would simply like a taste of the good life. Total revolution though, means the execution of the mysterious Mr. Wilford (Ed Harris), who designed and runs the train, as well as any other elites who have kept the passengers in their place.

Another important facet of Curtis is his ability to work through and see the mechanisms of hegemonic control. Shortly before they break out of the rear car, Curtis and Gilliam observe the guards. Curtis realizes that the guns have no bullets and are simply for show: "I think the guns are literally useless. They used up all their bullets four years ago in the last revolt. Bullets are extinct." The display of an assault rifle signals the possibility of violence, which in this case has been enough of a coercive state apparatus to keep people at the back of the train.

Curtis leads the escape out of the rear car and slowly past the guards. As the revolutionaries get into each successive car, both them as characters and us as viewers learn more about the technical and social aspects of the train. Perhaps most interesting is how Wilford is deified by the passengers. His story is told to children in an elementary school car in yet another display of hegemonic control. The children learn of Wilford's Thomas Edison like qualities of grit, determination, and vision. Because Wilford has created such a perfect home for these children in the cold, dead planet, anyone who criticizes his design is an enemy. Possible holes in this ideology are quickly stitched up through a call to nature. The suffering of those at the back of the train is taught as a self-imposed suffering due to the natural laziness and immoral behavior of those passengers.

It is a violent path to the engine at the front and most of the revolutionaries are killed on the way. Eventually, Curtis makes it to the front and is invited in to meet Mr. Wilford. It is here in the very front that everything is explained to Curtis. Wilford states that the entire train is an ecosystem that must have balance. If the population gets to large, Wilford explains, invoking Malthus, it needs "to be reduced rather . . . drastically."[24] There is no time for natural selection in this ecosystem; all would suffer as resources diminish. He explains to Curtis that his and previous revolutions and riots have all been caused by Wilford in an effort to thin the herd through the resulting violence. The only reason Curtis was allowed to get to the engine was because Wilford wished it. In fact, Wilford and the now deceased Gilliam were working together to facilitate the whole revolution; they simply didn't expect Curtis to get this far. The constant presence of anxiety and fear help to maintain a social balance as well, which these revolutions also help maintain.

Curtis is offered the commanding position at the front as the aging Wilford gets ready to retire. Wilford seduces him with comfort, power, and knowledge and Curtis clearly is unsure as to whether he should stick to his revolutionary principles or embrace a more comfortable position within the status quo. He is clearly leaning toward the latter when it is revealed that the children taken from the tail section are selected for their small size in order to keep the train engine working. This is not in the sense of working as technicians but as actual mechanical components of the engine itself. Young Timmy (Marcanthonee Reis), one of the boys we see kidnapped early on in the film, is awkwardly folded into a small compartment preforming rhythmic movements to allow the train to keep running smoothly (figure 3.3). Parts are not easy to come by in this new reality, but children are. Seeing the children suffer within the engine is enough to help him make a decision; Curtis sacrifices his arm to free Timmy. The film ends with the train crashing off the track and into the snow. Presumably the only survivors are a young girl, Yona (Ko

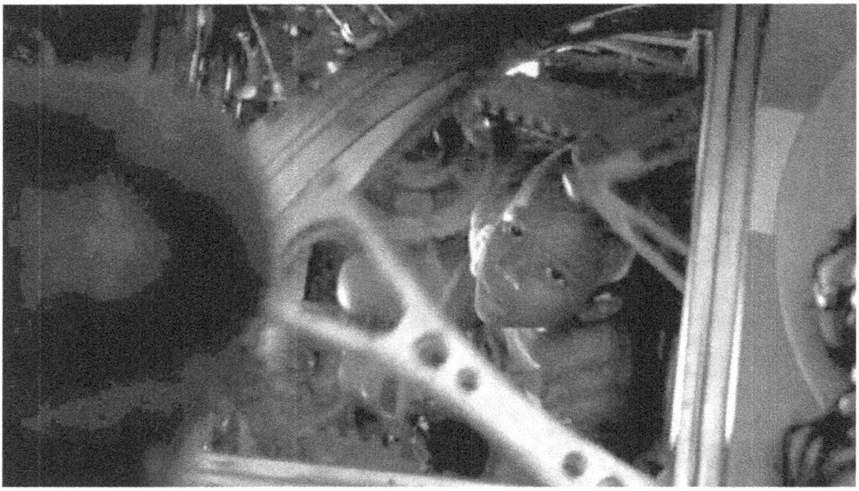

Figure 3.3 Timmy, Like Other Small children, Replaces Worn Out parts in Snowpiercer's Engine. *Source*: *Snowpiercer,* dir. Bong Joon-ho (2013)

Asung) and the kidnapped boy, Timmy who leave the train and venture into the cold unknown. The last shot is of a polar bear, letting us know that life, and therefore hope for our survivors, is not completely gone from the planet.

THESE AREN'T THE DROIDS YOU'RE LOOKING FOR

The three films, *Pacific Rim*, *I Am Mother*, and *Snowpiercer* work together to reveal our conscious and unconscious anxieties and hopes surrounding the climate crisis. All involve technology, though they do so in differing ways. *Pacific Rim* takes the technologically whiggish approach; we just need better machines dammit! We can punch our way out of this if we simply work together with our fellow humans. *I Am Mother* asks about the ethics of dealing with the climate crisis. While it acknowledges capitalism's role in getting us into the mess, the film simply argues that we need better people in charge of economic and social reproduction to ensure a logical path forward. We just need better people dammit!

Snowpiercer is the only film that doesn't try to show technology as the way forward. Machines cannot think, but people can and, short of a robot extinction event, people will continue to separate along class lines no matter what the Earth looks like. The film works in the Marxist tradition to question and critique what happens when a select few hold the means of production. The scene in which Curtis discovers Timmy in the engine reveals not only the cruelty possible in utilitarian philosophy, but simultaneously illuminates the labor inherent in the technology meant to save us from the climate crisis.

The machine precedes the climate. A film like *Pacific Rim* spends more time on the end of the world than it does questioning a character like Hannibal Chau. When one steps back from film and looks at the current options to escape the climate crisis, it is clear that we have a *Pacific Rim* mentality. The rather vaguely worded 2015 Paris Agreement, the latest international effort to battle the Kaiju of greenhouse gases, explicitly states in Article 2 the goal of "making finance flows consistent with a pathway towards low greenhouse gas emissions and climate-resilient development."[25] Later in the document, Article 6, Paragraph 8, acknowledges the "importance of integrated, holistic, and balanced non-market approaches" but only alongside things like finance and technology. Article 10 is specifically focused on technology: "Parties share a long-term vision on the importance of fully realizing technology development and transfer in order to improve resilience to climate change and to reduce greenhouse gas emissions." The very notion that new technology is the only way forward is never questioned, but instead rendered natural. The machine precedes the climate.

The Paris Agreement does aim to bring happiness to the masses through poverty eradication. And yet, no article of the agreement ever gets into just how poverty will be eradicated, nor questions the connection of class, capital, and climate. Rather than pull the children out of the machine, we are maintaining the engine of modern society no matter the cost. What other choice do we have within the simulacrum?

The films discussed here should be read against material conditions of both society and global atmospheric conditions. In doing so, it becomes clear that despite noble intentions, a film like *Pacific Rim* or *I Am Mother* is not representing the real but instead actively making simulacra. Only *Snowpiercer*, with its materialist storytelling, breaks the illusion to show the child in the machine and to conceive of burning it all down to produce something genuinely new. This materialist approach is the only way I can see to escape the Baudrillardian desert of the real and to live in nature that is no longer hyperreal.

NOTES

1. Jedediah Purdy, *After Nature: A Politics for the Anthropocene* (Cambridge: Harvard University Press, 2015), 5; Jason W. Moore "The Capitalocene, Part I: On the Nature and Origins of Our Ecological Crisis," *The Journal of Peasant Studies* 44, no. 3 (2017): 596; Bruno Latour, *Facing Gaia: Eight Lectures on the New Climatic Regime* (Cambridge: Polity Press, 2017).

2. Tim Dant, "The Driver-car," *Theory, Culture, & Society* 21, no. 4–5 (2004): 61.
3. David Harvey, *A Companion to Marx's Capital: The Complete Edition* (London: Verso, 2018 [2010]); Karl Marx, *Capital: A Critique of Political Economy, Volume One* (London: Penguin Books, 1990 [1867]).
4. Marx, *Capital, Vol. 1*, 493.
5. Harvey, *Companion*, 194.
6. Harvey, *Companion*, 195, his emphasis.
7. Harvey, *Companion*, 198.
8. Gilles Deleuze and Félix Guattari, *A Thousand Plateaus: Capitalism and Schizophrenia* (Minneapolis: University of Minnesota Press, 1987 [1980]), 25.
9. Jean Baudrillard, *Simulacra and Simulation* (Ann Arbor: The University of Michigan Press, 1994 [1981]), 1.
10. Jorge Luis Borges, "On Exactitude in Science," in *Jorge Luis Borges: Collected Fictions* (New York: Penguin Books, 1998 [1946]), 325.
11. Baudrillard, *Simulacra*, 6.
12. Baudrillard, *Simulacra*, 47.
13. Jean Baudrillard, *The Gulf War Did Not Take Place* (Sydney: Power Publications, 2009 [1991]).
14. Mark Fisher, *Capitalist Realism: Is There No Alternative?* (Winchester, UK: Zero Books, 2009).
15. Baudrillard, *Simulacra*, 21, his emphasis.
16. Tom Russo, "'Pacific Rim' is the heartfelt project from del Toro" *Boston Globe*, July 6, 2013, https://www.bostonglobe.com/arts/movies/2013/07/06/pacific-rim-latest-heartfelt-project-from-guillermo-del-toro/aTU5X0wLpNaUe1GpbHqDUO/story.html.
17. Dant, "Driver-car," 61.
18. Dant, "Driver-car," 74.
19. What is interesting is how such a wall will most likely how developed nations respond to sea-level rise and increased hurricane activity.
20. Jeremy Bentham, "From *an Introduction to the Principles of Morals and Legislation*," in *Utilitarianism and Other Essays* (London: Penguin Books, 2004), 65.
21. Antonio Gramsci, *Selections from the Prison Notebooks*, edited and translated by Quintin Hoare and Geoffrey Nowell Smith (New York: International Publishers, 1971), 348.
22. Slavoj Žižek, *The Sublime Object of Ideology* (London: Verso, 2008 [1989]), 12.
23. Gramsci, *Selections*, 10.
24. Unlike the favorable read of Bentham in *I Am Mother*, this Malthusian fear of overpopulation is presented as horrific.
25. United Nations, *Paris Agreement*, 2015, https://unfccc.int/files/essential_background/convention/application/pdf/english_paris_agreement.pdf.

Chapter 4

Zombies and the Horror of Not Having a Car

Apocalyptic Stories as Ecomobilities

The Anthropocene, if we accept it as an epoch of human/nature connections, offers new possibilities for how humans produce nature in the Marxist sense.[1] At the same time, everything we know about the changing climate suggests that this new epoch will be horrific for most humans. Once in a lifetime weather disasters are now common occurrences. How can we get excited about new connections within the environment, new metabolic interactions, if we live in constant fear of either drowning or burning to death?[2]

I started writing this chapter in March 2020. The world was in the beginnings of the COVID-19 crisis, a coronavirus pandemic that has crippled the global economy and arrested the movement of humans everywhere. Terms like "flattening the curve"—referring to slowing the rate of infection over time—and "social distancing"—encouraging six feet of distance between individuals and a limiting of large gatherings—are now commonplace. Listening and watching a variety of media throughout this time has revealed a line of thought that this pandemic is just one of many that will come as climate change increases. As humans encroach on nature, the thought goes, nature will fight back. And while it is not fully clear if COVID-19 is a product of the Anthropocene or simply a bad disease, clearly environmental degradation and a destruction of species' habitats make new opportunities for disease to spread.[3] If nothing else, we have definitely seen that the class inequalities across the globe have made the world ill prepared for any future crises.[4]

I want to use this chapter to explore the connections between the horrors of the Anthropocene, be they of the atmosphere, biosphere, lithosphere, or hydrosphere, and automobilities. I use horror deliberately here, both as a genre of film, but also as an affective response to external forces. The horrors associated with the Anthropocene have been massive in scale, initially confined to ecological degradation and natural disasters, but now including deadly pandemics too. While not every film about an apocalypse is classified as a horror film, the terror produced by the collapse of society plays on the same themes. Erin Y. Huang defines horror as "a socially produced affect that responds to contemporaneous forms of violence and that is basically antirepresentational but requires a form of representation."[5] The horrific has been studied at length when it comes to the body,[6] but Huang focuses on what she sees as second site, that of viewing horror "as a historical mode of perception arising when the perceived external reality exceeds one's internal frame of comprehension."[7] In other words, horror films (and I would include the apocalyptic genre) provide a way to take some existing abstract unease and violence and represent them as oil wars, geophysical disasters, or zombie outbreaks. Regardless of the cause, the effect is the anxiety of familiar systems failing coupled with a material violence of highway bandits or undead predators. The characters respond to these representations of antirepresentational violence by either explicitly or implicitly evoking key themes that help at least some of the protagonists make it to the end of the film. Repeatedly, filmic depictions of apocalyptic horror explore the ideas of habitat, consanguinity, and biological interactions within an ecosystem. Such themes of family and home are not new to humanity's story-telling and yet, the ideology of American automobility works to naturalize the connection between the driver-car assemblage and surviving long enough to keep family alive. The assemblage is extended to not just include the driver, but the driver's family. The family must move quickly across a landscape to survive. Stasis means certain death.

YOU CAN'T DRIVE AWAY FROM GERMS: HORROR, PANDEMICS, AND ASSEMBLAGES IN THE APOCALYPSE

In the United States, the response to COVID-19 was slow, causing the country to quickly race to the top of the charts in infections and death. While this is mere speculation, I cannot help but think that our failure as a nation here is due to our incessant need to move colliding with a pandemic that demanded we stay home. Deleuze always understood this about the country, explaining it through Anglo-American literature: "American literature operates according to geographical lines: the flight towards the West, the discovery that the true East is in the West, the sense of the frontiers as something to cross, to push back, to go beyond. The becoming is geographical. There is no equivalent in France."[8] Deleuze was explaining this idea as he and Guattari were developing *A Thousand Plateaus: Capitalism and Schizophrenia*, which not surprisingly becomes part of this later work. American literature is connected to movement, not just for the sake of moving, but to exist rhizomatically, horizontally, to not get bogged down by useless origins or hierarchies, but to become something new in the world.[9] To move rhizomatically is to produce "smooth" spaces of "deterritorialization" as opposed to the "striated" spaces of the State, that which fixes the vertical hierarchies of which Deleuze and Guattari attack.[10] While perhaps not invoking Deleuze, I have found that many of my neighbors in this rural patch of Northern Los Angeles County despise the orders to stay still, to allow a State territorialization to occur as they are locked in their homes. Obviously, coronavirus germs can exist in both smooth and striated spaces, but I cannot help but think that America's embarrassingly poor response to the pandemic is tied to concept of American mobilities.

Regardless of how Deleuze or Guattari use the term in discussions of the mobilities of literature, I want to use "America" as discussed earlier in the introduction of this book, borrowing from Seiler, as a way to "signify myth, transmit ideology, and confer power."[11] I am not so much interested in materially travelling across the place of the United States as I am in the concept ofAmerican mobilities that exists globally during the Anthropocene. I see a connection between mobility and the climate crisis in our response to horror that has yet to be explored. To move, in a rather specific way as I will show in what follows, is to try to take some control over antirepresentational violence, even if the effort is futile. Again, we can return to Deleuze to make sense of this American tradition of mobility: "The great and only error [lies] in thinking that a line of flight consists in fleeing from life; the flight into the imaginary, or into art. On the contrary, to flee is to produce the real, to create life, to find a weapon."[12] For Deleuze, and Guattari, lines of flight, assemblages, and rhizomes, all get at this concept of spreading outward to create something new. A book like *A Thousand Plateaus* is rife with examples of mobilities, but we should not assume that movement is the same as rhizomatic becoming. Rather, the ideology of American automobility, as I have been defining it throughout this work, is an assemblage of humans, machines, landscapes, and ideas. Only through the interaction of these different things can a driver (and a pedestrian and one who is immobile, and, and, and . . .) go through this act of becoming. The automobility is perceived as the mechanism that is "enabling one to blow apart strata, cut roots, and make new connections."[13] And a key component of this becoming is precisely American automobility's individual nature.

While the years 2020 and 2021 felt rather apocalyptic to many Americans in the closing and stopping of society, our usual approach to the end of the world as depicted in art has involved automobility of some kind to escape the horror. The films of the previous chapter question how a group or governing force responds to a threat to society. This chapter will look at how individuals deal with a violent shock at the beginning of a global crisis.

The American approach toward the apocalypse is that of driven individual success. Getting a handle on a pandemic like COVID-19 requires collective, and often passive, action in the sense that we are told to stay home. Interestingly, our response to climate change in general should look the same. The energy demanded by continued capital accumulation has infected the climate. Slowing down and focusing on the immediately local would limit emissions and go a long way toward staying under a 1.5°C increase from pre-industrial levels.

And yet, while our ideology has long prepared us for a specific type of disaster, that disaster is one that we are not currently facing, however. It is within the context of Americans stumbling through COVID-19 that I want to spend this chapter exploring the ideology of American automobility within stories of the apocalypse, zombie induced or otherwise, and how such stories connect to environmental change.

DRIVING AWAY FROM ZOMBIES

George Romero's *Night of the Living Dead*, the film that started the zombie genre, is laughably invested in automobility. This does not take away from the enjoyment of the film; the ideology has already been rendered natural so as to make it invisible to most viewers. I say laughable though, in the sense that the zombies are so slow and lumbering that a car hardly seems necessary to escape them. Give any of the human characters a halfway decent bicycle and they could have easily escaped.

The film opens with an empty, winding, country road. About a minute in, a car finally drives into the shot, siblings Johnny and Barbara are headed to pay respects at a cemetery when they encounter the first of many zombies. Johnny is killed, but Barbara is able to escape, not by running away, but by coasting downhill in the car. Johnny had the keys, seemingly rendering the car useless, but Barbara can still use the wheels and gravity to flee.

Night of the Living Dead also introduces the complexity between the home as a space of safety and horror. The various human characters converge on a farmhouse as a sanctuary, but the zombies simply continue to work to break in. It is clear that the characters must flee the home, though their quest for automobility will kill a few as they try to fuel up a truck. Ben, the final survivor of the night in the house meets his end as a militia spots his movement in the house and mistakes him for a zombie. If only he had kept moving and resisted the roots of the home.

Later films would change the zombies into faster killing machines, thus making an automobile an actual necessity. Ruben Fleischer's 2009 film *Zombieland* goes so far as to stress the need for buckled seatbelts when escaping zombies to avoid getting thrown through the windshield if the drive gets rough. The whole film is a homage to American automobility, though in such an overt way that it does little to advance the common sense of the ideology. Instead, the film relies on the ironic distance of a Žižekian ideology, we are meant to acknowledge the absurdity of the literal ideology (a bright yellow Hummer H2 with the late Dale Earnhardt's Number 3 spray painted on the side) while nonetheless reproducing it.[14] The characters Columbus (Jesses Eisenberg), Tallahassee (Woody Harrelson), Wichita (Emma Stone), and Little Rock (Abigail Breslin) all have adopted their names as a signal to their origins, and yet they must continue to move on the roads to avoid infected humans. The film takes the characters on Route 66 on their journey to California all the while revealing that they have lost other family members to the zombie outbreak. Of course these characters will become a new family, facilitated by the long hours in the car as they head west. The road trip is the social practice that produces the familial unit. It is made absurd with the addition of zombies, and yet, the actual practices are held up as normal: "For fuck's sake, enough already! We are being chased by ravenous freaks! We don't have enough problems . . .? We can't just fucking drive down the road playing I Spy or some shit for two hours like four normal-ass Americans?"

Ultimately, the ideology of American automobility is celebrated in *Zombieland*, but it is in another zombie film, Danny Boyle's *28 Days Later*, that the automobile acts as another character, one crucial to the story though never being fully revealed.

The importance of *28 Days Later* is its naturalization of the car-human assemblage and the connection of family to driving. Much like *Mad Max: Fury Road* (see chapter 2), the characters must assemble with automobiles in order to survive, but the assemblage is only necessary for a certain type of becoming in this zombie-infested wasteland. Characters also have the option to reject automobility and to instead embrace a sedentary dwelling, though our protagonists will go to great lengths to "cut roots."[15] Movement is important, but in *28 Days Later*, it is not enough. The driver-car assemblage, in which "neither the human driver nor the car acting apart could bring about the types of action that the assemblage can" is crucial to escape both infected humans wanting to kill for killing's sake as well as the non-infected who kill for insidious reasons.[16]

28 Days Later begins with Jim (Cillian Murphy) exiting a hospital after waking to an empty building, only to discover that all of London is empty. The montage of shots that follow are disturbing in their absence of mobility. While Jim walks, the streets of London are empty. Vehicles and their drivers should be buzzing about the roads, but everything is completely frozen in place. The city appears to be dead. The effect is haunting, though I cannot help but connect this to the COVID-19 pandemic. Shouldn't our cities all look like this as we combat a virus? What are the ramifications for society of the horror invoked by seeing a still city?

It soon becomes clear to Jim that the city isn't so much dead as it is infested with twitching rage zombies that want nothing more than to attack uninfected humans. A small group of survivors hide throughout London, two of which, Mark (Noah Huntley) and Selena (Naomie Harris), find Jim and explain what has happened. Jim insists on going to check on his parents, despite Mark and Selena's assurances that they are most likely already dead. They go to the house nonetheless, and sure enough, his parents have committed suicide, likely what they saw as the only way to truly escape this horror.

As if finding his parents dead wasn't traumatic enough, zombies break through windows shortly after the three arrive. This childhood house, a dendritic root system to stick with our Deleuze and Guattari speak, offers nothing for him.[17] His only option is to flee, not to escape, so much as to become something new, a new assemblage. Selena too represents this break from roots and strata. After they have killed the zombies in Jim's childhood home, the three pause to survey the carnage. Mark's arm is bleeding, presumably from a zombie bite. Without missing a beat, Selena violently hacks him to death. We don't know much about their backstory, but these two had been surviving together for at least a week or two, and yet Selena clearly understands the dangers of putting down roots. When Jim questions her about killing Mark, Selena explains:

> Look, if someone gets infected, you've got between 10 and 20 seconds to kill them. It might be your brother or your sister or your oldest friend. It makes no difference. And just so you know where you stand, if it happens to you, I'll do it in a heartbeat. He was full of plans. Do you have any plans, Jim? Do you want us to find a cure and save the world or just fall in love and fuck? Plans are pointless. Staying alive's as good as it gets.

Selena clearly has been scarred by the zombie epidemic, but her acceptance of the need to break from London (a city full of roots and trees) is the catalyst for the rest of the film even if Jim is the one who suggests leaving.

Shortly after Mark is killed, Selena and Jim stumble upon two other survivors hiding in an apartment, Frank (Brendan Gleeson) and his daughter Hannah (Megan Burns). Frank has discovered a recorded radio broadcast that announces that there are soldiers waiting to help survivors at a site 27 miles northeast of Manchester. They claim to have "the answer to infection" without giving any details. While the journey (surely to be full of zombie encounters) is frightening to consider, supplies in Frank and Hannah's apartment are diminishing. Mobility is the only answer, and specifically one that takes them away from urban areas. As Hannah explains, "We'll never be safe in the cities."

The four load up in Frank's taxicab, a small, bulbous LTI Fairway that does not convey the image of a rugged zombie-crushing rig (Figure 4.1). Nonetheless, the car represents a shift in the film, one in which our protagonists might have a shot of surviving this new world. Sound is important in producing the landscapes of the film. At times the scenes are punctuated with aggressive rock, while other scenes are devoid of any non-diegetic sound to produce an eerie emptiness. This scene is notable for the beautifully ethereal singing that brings liminality to the inside of the car. Frank acknowledges the meter of the cab and makes a joke about not taking credit cards. The car is safe, or more specifically, the inside of the car is safe from the outside apocalyptic horrors. What Danny Boyle has done with the specific montage and soundscapes is to relax the viewer when the taxi is moving. Mobility means safety from danger, which when you think about it, is absurd. The very act of buckling a seatbelt is a reminder of just how violent driving can be, and yet, the viewer accepts the safety of Frank's taxi.

To leave London, the four must drive through a tunnel. As they drive deeper into it, dead bodies appear followed by a pile of wrecked cars from people desperate to leave at the start of the outbreak. Rather than turn back, Frank deftly drives over the massive heap of metal, glass, and rubber. The ethereal music returns and drowns out the awful noise of the taxi rolling over the wrecked vehicles. The four laugh at the absurdity of the moment while also believing that they have made it. A popped front tire kills the mood.

Figure 4.1 Frank's Cab Loaded Up to Make the Zombie-Filled Journey in *28 Days Later*.
Source: *28 Days Later*, dir. Danny Boyle (2002)

The four hop out of the car to change the tire as quickly as possible. Hannah pulls the jack under the car. Jim watches for zombies and hears a noise growing in the distance. Thousands of rats are racing toward the car, which means poor Hannah is quickly engulfed by them. Hannah is fine, but Selena realizes the rats are running from the infected humans racing that way. Our protagonists race to fix the tire with frantic cuts of running zombies, panicked humans, and Frank lifting the car to help speed along the repair. They fix it just in time and speed away from the danger. They are inside the moving car; they are safe.

This scene marks the end of their encounters with urban spaces. The four stop at a grocery store, but it is in an empty suburb. Apart from Jim's baseball bat, there is no evidence of the group being truly concerned about an attack here. The city was full of danger, but that is behind them. They are carefree as the four of them each take a cart to load with supplies for their journey.

The taxi takes the group further from the city and into the rural countryside. They must stop for fuel, which requires surveillance of the area. Jim discovers the decaying corpses of a family as well as a terrifying boy zombie, but honestly, the scene is not as tense as anything back in London. They continue driving on an empty multilane highway, ethereal music playing of course, and eventually stop at the ruins of an old country estate. In any other film, the moss covered stone walls would normally be a perfect setting for a zombie attack, but instead the four stop to have a relaxing picnic. They sit in a green field joyously devouring food. Frank surveys the area, not out of fear, but to take in the beauty of rural England. He calls to the other three to come look at four horses running in a field. "Like a family," says Frank, while the camera is not on the horses, but rather framing the four of them like a family portrait. Where London took their previous biological families and homes, their automobility has facilitated a Deleuze/Guattarian becoming. The assemblage of humans, machines, horses, nature has produced something new for our protagonists.

Frank makes the call for them to sleep at this spot for the night. Again, the stone ruins would be a great setting for a zombie attack, but this space is a safe and familial one. That night, with the help of Xanax, Jim sleeps and has a nightmare in which he wakes up alone, abandoned by his new family. As he mutters in his sleep, Frank walks over and gently reassures him. "Thanks, dad," says Jim.

In the morning they continue toward Manchester. This section of the film is simply shots of the taxi driving along empty highways overlaid with more beautiful choir music and sounds of nature. Jim holds his arm out the window to feel the wind. Hannah and Selena play cards. As they reach Manchester, the camera tilts up to show that the entire city is on fire which they will have to bypass to reach the blockade from the radio broadcast. The music is gone.

The camera shots all get tighter. The walls of urban living close in on the four travelers. The blockade is empty making them question if the trip was worth it. Frank is furious since he saw this as their only hope of survival. As he wanders around the empty space, a drip of blood falls from a suspended corpse causing him to change. As Jim gets ready to kill him with his bat, masked soldiers appear and shoot Frank dead. The soldiers take Selena, Hannah, and Jim back to a large estate that has been fortified with a watch tower, flood lights, barbed wire, and landmines. The family is violently shattered as soon as the car has stopped moving.

Major Henry West (Christopher Eccleston) greets them and offers hot water and beds with clean sheets. As Jim showers, several of the soldiers are goofing off while driving Frank's taxi in circles, clearly tainting the machine that brought the nomadic family such joy just moments ago. Major West later explains to Jim that the answer to infection is not a cure but to rebuild society from this house. As Jim gets a tour of the facility, the camera angle is off just a bit to let the viewer know that despite the domesticity implied by such a house, it is not safe. This space is one of urban dangers threatening to further wreck the family.

At a candlelit dinner around a large table, all of the soldiers and Jim, Selena, and Hannah dine on a variety of canned foods. They debate what the future holds and if a return to normalcy is possible, with Sergeant Farrell (Stuart McQuarrie) suggesting that when viewed geologically, humans have only been around briefly so an extinction would actually be a return to normal. He is mocked by Major West, who argues that human violence is the real normalcy, foreshadowing his reason for trying to attract travelers like Jim, Selena, and Hannah. The dinner is interrupted by a large group of zombies trying to breach the estate who are quickly dispatched by the machine guns and landmines.

We will soon learn that whole point of the radio broadcast was to lure females to the house so that the soldiers had the ability to breed and repopulate the Earth. Major West wants a future, one that can only come by setting down roots at the estate, even if means kidnapping and raping women to get it. Farrell, our philosopher sergeant is against the plan, so he and Jim are locked away to be executed. Farrell explains to Jim that the only logical explanation is that the disease couldn't cross the ocean to infect the rest of the world. The British Isles are quarantined and all they need to do is wait it out. This is confirmed when Jim escapes his captors and sees a jet flying high overhead.

Quite simply, Jim will come back and kill the soldiers to save the women (with the help of some zombies). The three make it back to the taxi (despite the presence of much more capable military trucks and jeeps around the estate) and find Major West hidden in the back. He shoots Jim in the stomach while Hannah quickly starts the car and backs it over near an infected soldier

Figure 4.2 The Car-as-Home Taxi Ready to Flee the Soldiers' Estate in *28 Days Later*.
Source: *29* Days Later, dir. Danny Boyle (2002)

who quickly pulls Major West out of the backseat. Hannah picks up Selena and the wounded Jim, the camera shoots this from within the house, framing the taxi with its columns (figure 4.2). The automobile, getting ready to move once again, is the true space of family, safety, and becoming. Just as the roots are closing in, Jim, Selena, and Hannah are able to flee into this assemblage and flatten "all of its dimensions onto a single plane of consistency."[18] They are nomads, which is the only way to stay safe.

The final scene cuts to the three survivors somewhere in the countryside. Jim is bandaged up and Selena and Hannah make a giant sign to attract the attention of a plane from the continent scouting the country for survivors. We can assume they are rescued soon after. Cut to black.

This final scene, one that actually implies roots as working out, was made later when the original ending was deemed too depressing for audiences. The original end, which can be seen if one sticks through the credits of *28 Days Later*, had Hannah and Selena pushing the wounded Jim on a gurney into an abandoned hospital (much like the one we first meet him in). Selena frantically moves to keep him alive through medication and CPR, but ultimately Jim dies. Hannah asks Selena, "What are we gonna do now?"

"We move."

MEN WILL BE MEN: *HOW IT ENDS*

The zombie films discussed above were all well received and commercial successes. But what of the made-for-streaming schlock that plays with similar themes? The 2018 film *How It Ends*, directed by David M. Rosenthal, is another tale of horror and the apocalypse, though this time with the absence of zombies. In this case, the unease and anxiety come from a collapsed communications network and electrical grid in the United States after a mysterious disaster that appears to be natural but might be a military action from a foreign government. Much like *28 Days Later*, the film explores ideas of mobilities and family, but rather than Danny Boyle's critique of masculinity through the soldiers, *How It Ends* promotes a rather conservative concept of mobile American masculinity as *the* means to protect family. Additionally, where *28 Days Later* shows normal people doing their best in a bad situation, *How It Ends* is a doomsday prepper's dream. Preparation and combat training, two clearly masculine traits, are the key to survival in this apocalyptic hellscape.

The best treatment of the term "masculinity" is in the book *Female Masculinity*, in which Jack Halberstam refuses to assume the link of masculinity with the biologically male body. By treating masculinity as a performance, in much the same vein of Judith Butler's *Gender Trouble*, Halberstam is able to get at just how masculinity is produced.[19] *Female Masculinity* as a study is indifferent "to the whiteness of the male and the masculinity of the white male and the project of naming his power."[20] Halberstam is more interested in the study of those performances counterposed to a dominant white masculinity that dominates action films. Despite the rather banal treatment of masculinity in *How It Ends*, there is still something to be taken from how it is constructed on the road.

Will (Theo James) leaves his pregnant fiancée, Sam (Kat Graham) in Seattle to fly to Chicago for both business and to see her parents and ask her father, Tom (Forest Whittaker) for her hand in marriage. Will goes to the parents' apartment for dinner and we immediately learn that Tom is clearly a successful man based on his nice apartment with a gorgeous view and nice clothes, but he is also a gruff father who is clearly protective of his daughter. Despite Tom being an African American, he fits what Halberstam sees as the dominant conception of American masculinity, which apart from being distinctly white also

> inevitably conjures up notions of power and legitimacy and privilege; it often symbolically refers to the power of the state and to uneven distributions of wealth. Masculinity seems to extend outward into patriarchy and inward into the family; masculinity represents the power of inheritance, the consequences of the traffic in women, and the promise of social privilege.[21]

As will become clear as the film develops, Tom will fit all of these criteria. And despite Forrest Whitaker's great performance, one of the few pleasurable things about the film, the character could have easily been played by a white actor. His being Black never comes up, even out on the road when being accosted by white and heavily armed militias. The absence of race for the character Tom further solidifies his position of possessing a dominant form of American masculinity.

Tom clearly does not approve of Will's presence in his daughter's life. One issue is Will's apparent lack of a long term plan to become financially stable, and therefore be able to support his daughter. Tom explains it to Will during the argument that will ultimately put an early end to the dinner: "Everyone wants the dream. The problem with your generation is nobody wants to work for it. I spent 27 years serving our country. I saved and I saved until I started working for Northbridge, where I started making some real money. But I always had a plan." Tom is not including his wife Paula (Nicole Ari Parker) in this statement; he evidently believes in a singular, masculine provider within any heterosexual relationship. He is happy to take care of Paula, but Tom clearly makes the decisions.

The film's plot quickly gets underway as a "large seismic event" in Southern California causes power outages across the Western United States and soon affects the rest of the country. Airports are useless. Something massive is clearly wrong, yet both the characters and the audience have no idea what exactly is happening. Tom is not impressed with Will's indecisiveness as the events unfold: "Let's look at what we know, Will. There was an event, a couple of hours ago, out West. The power shut off here, two thousand miles away. We have no idea what's happening, yet we've got F-22s doing flybys. This moment is not about waiting for the power to come back on. The only thing that we can control is what we decide to do." Will asks him what he is going to do, to which Tom replies, "Get on the road." There is a beat to let that idea sink in and then asks, effectively, if Will is man enough to join Tom on an apocalyptic road trip to Seattle to save his, what he assumes, helpless and scared daughter. Paula will spend this time with their son nearby, obviously. Before Paula drives off, Tom reassures her that their son is "prepared for everything" and therefore she will be safe. They say goodbye to Paula and the two men drive off in Tom's Cadillac.

How It Ends plays up what happens to the American road trip when the system fails. Rather than play up the absurdity as in *Zombieland*, *How It Ends* uses the apocalyptic horror to reveal how desperately our society needs automobility and the ability to travel large swaths of the country. The romantic image of crossing large sections of the continent by automobile were promoted since commercial automobiles were first sold in this country. Peter Blodgett's collection of motor touring stories from the early twentieth century, for example, show the hardships in crossing the country, but invoke the romance of the frontier rather than apocalyptic despair.[22] This film, however, presents the American road trip not as a rewarding adventure, but as suffering in the absence of modern convenience. Before Tom and Will really begin to drive, they fuel up at a congested gas station. Handmade signs announce that transactions can only be made in cash. Motorists appear stranded and frustrated. Tom goes into the food mart for supplies and has Will fill up the car and a jerry can for their journey, which shows just how prepared Tom is. Others are stranded at the gas station; Tom has taken the steps to make sure he can make it to Seattle. While this journey will be difficult, men with the proper planning and equipment can succeed. Further, when a few people try to get money, if not something more, from Will, Tom pulls out a handgun, which sends the would-be muggers running. Will is furious that Tom has a gun in the first place, but Tom argues that if he didn't have the gun, the car would probably be gone and their hopes of saving Sam would be gone. Will cannot dispute this.

The gas station scene is the first of several that show society collapsing, but it is the most interesting precisely because it happens so shortly after the power outages occur. While pumping gas without a credit card can be frustrating, the despair shown by the motorists does not seem warranted. The same goes for the muggers so brazenly threatening violence in the morning light with dozens of witnesses. What *How It Ends* manages to do, is to naturalize the idea that our electronic conveniences, especially those connected to our automobilities, are the only thing keeping society functioning. Credit card readers on a gas pump and GPS navigation in our dashboards are the only things keeping us from devolving back into violent animals. A year ago, I might have really critiqued this idea as being completely absurd, but having seen stores with empty toilet paper shelves and rowdy customers at the start of the COVID-19 pandemic I know how little blips in the overall order of things can cause major changes. I shudder to think how close to reality *How It Ends'* hellscape actually is.

Most of the road trip takes place on westbound Interstate 90. As William Philpott says, "a highway is neither organic to its landscape nor inevitable in its origins" so it is no surprise that the film contains many aerial shots of the landscape cut in half by the Cadillac driving on the long straight road.[23] The absence of humans and community on the interstate plays into the recurring theme that any human outside of the car is likely a threat to Tom and Will. And with society failing all around them not even familiar institutions are safe. They are accosted by a fake state trooper on the very first night of their road trip in a violent scene that lets the viewer know that is that society is entering some sort of Mad Max hellscape on the very first day of a disaster. This scene and others to come on the road naturalize the idea that ultimately humans are a flawed, violent species that will prey on the weak without the presence of strong societal institutions. We are not meant to assume that humans could be out on the interstate to help fellow travelers. Further, Tom's

Marine Corps training and experience is what has allowed him to turn this immanent human violence into a productive force. He can help those who are helpless through his very competence, which is the ultimate form of masculinity in *How It Ends*.

Tom and Will wind up on the Crow Reservation in Montana. A group of elder indigenous men sit on a porch around a shortwave radio and bring up the voice of a man questioning what has happened. Suddenly a woman's voice on the radio screams, "Mayday, Mayday!" We cut to Tom's point of view as he gazes up into the sky. Thousands of birds are flying erratically. Everyone now is nervously looking to the sky. Familiar institutions are failing, but nature clearly knows that something is wrong. Up until this point, whatever disaster occurred has wrecked American society, but there was nothing to suggest an environmental catastrophe. The birds acting strange combined, unfortunately, with the introduction of the reservation is a way to connect the crisis to a more primal, natural issue. Being on Crow land, even if it was arbitrarily made their land by the US Government, allows the disaster to finally be seen as something being wrong with the Earth and not just society. The seemingly erratic flight of the birds suggests that nature knows something is amiss.

Tom will convince a mechanic, Ricki (Grace Dove), to come with them to Seattle so that they have someone who can help with any mechanical issues on the road. What is interesting is that a female character might disrupt the masculine dynamic in the car, but Ricki is not a feminine character. She has relatively short hair and is tough enough to talk back to an alpha male like Tom. We are meant to infer that she is leaving an unhealthy if not abusive heterosexual relationship on the reservation, but she presents an indigenous female masculinity throughout the film. Halberstam explains the original role of the "butch" in film: "She is tough and tragic, she was a tomboy, and she expresses a variety of masculinities."[24] The butch, Halberstam explains, also usually has a male sounding name. Ricki's gender is complex though, as she is able to fix a car, which the men cannot, and as we will find out later, she can shoot a gun, but at one point she strips to her underwear to reveal her female figure. The scene does not do much for the film's storyline, but is an odd juxtaposition of inviting the male gaze to the least feminine female character in the film. If the film were not so conservative in its presentation of gender roles I might suspect the filmmaker was playing with gender, but I doubt that is the case.

Ricki's female masculinity is complicated, but her role in the film is to be the indigenous connection to the Earth. This is an essentialized connection, one that assumes indigenous peoples have a special, sustainable ecological knowledge that is encoded within their DNA.[25] At one point in the trip, Ricki passes forward a magnetic compass to show that the needle is spinning wildly around. The fact that the geomagnetic North Pole either switched places or the laws of magnetism no longer work suggests that nature is behaving abnormally, but it is most telling that Ricki is the one who brings attention to this. Again, the indigenous person can read nature in a way that a European- or African Americans cannot. The characters don't have time to really assess the significance of the compass as two other cars appear out of nowhere and collide with one another. While one of the people involved in the accident is injured, but still alive, Tom argues there is nothing that can be done since they are miles away from any hospital. The three cannot stop moving or they will end up dead themselves. Ricki is clearly doesn't want to leave the survivor and her reaction to the situation reads as more emotional than that of the stoic, practical men. A car full of three armed men arrives at the scene of the accident, we can assume to steal or do worse. Our protagonists get away just in time. Everyone on the road must be treated as a potential enemy.

Much like in *Zombieland*, Tom, Will, and Ricki encounter familiar sights and events that a typical American road trip would contain. The difference, however, is not the silliness inherent to the characters and zombies of that film, but instead treat this new world (again only a day or two old) as a sad, predatory landscape. For example, the three later pull off the highway to an abandoned water slide park and diner. As they search for gas to siphon and collect some food from a broken vending machine, the camera frames the shots from behind structures to make us think they are being watched. This roadside attraction serves the same purpose as it did before the catastrophe (Ricki even jumps in a pool to cool off, only to find it is unusually hot), only in a much more ominous way.

Ricki has to leave the film at some point, because while her own masculinity is not at odds with Tom's or Will's, she gets in the way of a reproduction of distinctly male masculine roles. After a violent encounter with highway bandits, and the fascinating sight of an exploding hybrid Honda Insight hatchback, Ricki cannot take the brutally of this trip any longer. She walks off, away from the Cadillac and the road and shouts back to Tom, "We'll survive this longer than you!" I can only assume she is referring to her indigenous identity here, though these last words are cryptic. Tom and Will never see Ricki again.

By the time Ricki is gone, it is clear that Tom will soon die from injuries he sustained in their run in with the phony state trooper if he doesn't get to a hospital. With death near, the two men have a heart-to-heart as they continue down the interstate. "You know I love your daughter," Will says. "I know . . . If I didn't think that, I would have killed you a long time ago," Tom replies. Through bloody coughing fits, Tom makes Will promise he will keep his daughter safe. Will responds that of course he will, rather than begin any discussion of Sam's ability to take care of herself. By this point the viewer should readily accept that Sam needs a strong, competent masculinity in her life to complement her gentle femininity.

Tom dies after yet another violent encounter with even more Mad Max like villains (we're four days into the power outage). The Cadillac is also all but destroyed in the encounter. With somber music playing in the background, Will pours gasoline into the Cadillac, lights a road flare, and tosses it in to give Tom something resembling a Viking funeral along the interstate (Figure 4.3). Tom stares at the burning pyre which I cannot help but see as the pinnacle of American automobility, moving past an assemblage to a fully hybridized driver-car.[26]

Will now struggles to walk to Seattle along backroads to avoid the clear dangers of the interstate. Eventually a late 1980s Jeep Wagoneer pulls over, an SUV that is a powerful symbol of a past American automobility before fuel efficiency standards demanded smaller engines and smoother body lines.[27] The family not only gives him a ride but will give him the car since Will states he will need a four-wheel drive, though doesn't explain why he will need it. Perhaps it simply represents his masculine planning for the unknown.

As Will drives the Jeep closer to Seattle, the air grows red-gray and ash falls from the sky, looking like snow and evoking the 1980 eruption of Mount St. Helens. Will finds a gas mask in an abandoned fire truck, which he puts on to explore what is left of Seattle. It does not look good for his fiancé. Buildings are blown apart; the streets are covered in ash and debris. Fires burn everywhere. Where *28 Days Later* showed an eerily still London, Seattle in *How It Ends* is a disaster. Buildings are ruined, cars are flipped upside-down, and

Figure 4.3 Will (Theo James) Stands and Watches Tom's (Forest Whittaker) Funeral Pyre Burn in *How It Ends*. This is the pinnacle of American automobility. *Source*: *How It Ends*, dir. David M. Rosenthal (2018)

toxic ash covers everything. Will walks to his former apartment, which is really just a blown apart hull of a building, but luckily sees a message from Sam scrawled on a metal door that is rather untouched with an address that turns out to be a cabin in the woods outside of the city. Sam has been staying there with their neighbor, Jeremiah (Mark O'Brien), a creepy guy who clearly wants Sam to himself. After confronting Jeremiah about messing with the Jeep (never touch a man's Jeep), Will ultimately shoots Jeremiah who threatens to shoot first in a Western style standoff. Will's masculinity is complete as he protects Sam from Jeremiah.

Immediately as the shot is fired, the Earth begins to erupt and a giant pyroclastic cloud bears down on Will and Sam. They jump into the Jeep and start driving away. It doesn't look good. Will has Sam look him in the eyes and he tells her it will be okay. They both say "I love you" simultaneously. The Jeep seems to be pulling away from the cloud and the screen fades to black.

CAN WE OUTDRIVE THE HORROR?

Not only is American automobility a necessary means to escape horror, at least as far as the ideology would have you believe, but being on the road, with family, is a means to preserve consanguinity and habitat. And this ideology is precisely why COVID-19 has been so difficult for Americans. We envision our apocalyptic scenarios as something from which we can drive away. The "shelter in place" orders being given around the country force us to stay put, which frankly is much better than driving across a wasteland. But our ideology of a specific mobility is at odds with these orders.

The Anthropocene is full of horror thanks to the anxiety and violence that rising temperatures and increasing storms have brought. The unknown future of pandemics adds to the nightmare. All we want to do, we subjects of American automobility, is to drive away from these horrors, preserving family and a future along the way.

Horror films that deal not with individual monsters or haunted houses, but with a more apocalyptic wide-spread threat, reveal connections between our machines, societal relations, and the reproduction of daily life, as well as our connection to nature. What zombie films and the like show us is that, in a way much like the films discussed in chapter 3, we expect to be able to use our machines to either drive away from or into danger in order to solve problems. Zombies provide a visual representation of a clear threat that can be escaped, though directors like Ruben Fleischer and Danny Boyle are obviously using the premise to explore themes of family and roots. Automobility is naturalized through all of this so that Frank's taxi is a commonsense component of both family and escaping zombies. A more generalized natural apocalypse film like *How It Ends*, on the other hand, never tells us what exactly happened, though the title suggests that the disaster wound up being global and permanent. Family is also explored, though through masculinity being passed down from father to son, which again, is connected to automobilities although much more overtly. All of these films end with more mobility, we must move even if the odds don't look good. And in the midst of a global pandemic, I can see the draw in continuing to move. At least it's something to do.

In *Anti-Oedipus,* the first volume of *Capitalism and Schizophrenia*, Deleuze and Guattari discuss the lines of flight inherent to American literature, but do so in a way not quite as positive as in the second volume. "Isn't it the destiny of American literature that of crossing limits and frontiers, causing deterritorialized flows of desire to circulate, but also always making these flows transport fascisizing, moralizing, Puritan, and familialist territorialities?"[28] I cannot speak to why the men dropped the warning that mobilities can lead to both freeing and repressive spaces in later work, but the films discussed in this chapter are an excellent warning that mobilities do not only bring freedom or a positive becoming. To either romanticize the road trip or to ironically laugh at the addition of zombies to the practice is to fall into the ideology of American automobility and miss the fact that where the car goes, so too does the environmental degradation brought about by our capitalist system of mobilities. It is time to rethink our lines of flight, as they appear to bring the horrors along for the ride.

NOTES

1. Jason W. Moore, *Capitalism in the Web of Life: Ecology and the Accumulation of Capital* (London: Verso, 2015); Neil Smith, *Uneven Development: Nature, Capital, and the Production of Space* (Athens: University of Georgia Press, 2008).

2. Karl Marx, *Capital: A Critique of Political Economy, Volume One* (London: Penguin Books, 1990 [1867]), 290.

3. Aaron Bernstein, "Coronavirus, Climate Change, and the Environment: A Conversation on COVID-19 with Dr. Aaron Bernstein, Director of Harvard Chan C-CHANGE," *Harvard T.H. Chan School of Public Health*, https://www.hsph.harvard.edu/c-change/subtopics/coronavirus-and-climate-change/; It is worth mentioning that the Chinese Wet Market origin has never fully been proven, yet persists. See James T. Areddy, "China Rules Out Animal Market and Lab as Coronavirus Origin," *The Wall Street Journal*, May 26, 2020, https://www.wsj.com/articles/china-rules-out-animal-market-and-lab-as-coronavirus-origin-11590517508.

4. Shinichiro Asayama, Seita Emori, Masahiro Sugiyama, Fumiko Kasuga, Chiho Watanabe, "Are We Ignoring a Black Elephant in the Anthropocene? Climate Change and Global Pandemic as the Crisis in Health and Equality," *Sustainability Science* 16 (2021): 695–701.

5. Erin Y. Huang. *Urban Horror: Neoliberal Post-Socialism and the Limits of Visibility* (Durham: Duke University Press, 2020), 4.

6. Jack Halberstam, *Skin Shows: Gothic Horror and the Technology of Monsters* (Durham: Duke University Press, 1995).

7. Huang, *Urban Horror*, 6.

8. Gilles Deleuze and Claire Parnet, *Dialogues II* (New York: Columbia University Press, 1987 [1977]), 37.

9. Gilles Deleuze and Felix Guattari, *A Thousand Plateaus: Capitalism and Schizophrenia* (Minneapolis: University of Minnesota Press, 1987 [1980]), 19.

10. Deleuze and Guattari, *A Thousand Plateaus*, 353.

11. Cotten Sieler, *Republic of Drivers: A Cultural History of Automobility in America* (Chicago: The University of Chicago Press, 2008), 7.

12. Deleuze and Parnet, *Dialogues II*, 49.

13. Deleuze and Guattari, *A Thousand Plateaus*, 15.

14. Slavoj Žižek, *The Sublime Object of Ideology* (London: Verso 2008 [1989]), 12.

15. Deleuze and Guattari, *A Thousand Plateaus*, 15.

16. Tim Dant, "The Driver-car," *Theory, Culture & Society* 21, no. 4–5 (2004): 62.

17. Deleuze and Guattari, *A Thousand Plateaus*, 15.

18. Deleuze and Guattari, *A Thousand Plateaus*, 90.

19. Judith Butler, *Gender Trouble: Feminism and the Subversion of Identity* (New York: Routledge, 1990), 34.

20. Jack Halberstam, *Female Masculinity* (Durham: Duke University Press, 1998), 3.

21. Halberstam, *Female Masculinity*, 2.

22. Peter J. Blodgett (ed.), *Motoring West, Volume 1: Automobile Pioneers, 1900-1909* (Norman: University of Oklahoma Press, 2015), 19.

23. William Philpott, *Vacationland: Tourism and Environment in the Colorado High Country* (Seattle: University of Washington Press, 2013), 99.

24. Halberstam, *Female Masculinities*, 186.

25. Fikret Berkes, *Sacred Ecology, third edition* (New York: Routledge, 2012), 16.

26. Tim Dant, "The Driver-car," *Theory, Culture, & Society* 21, no. 4–5 (2004): 61.

27. Jeep Wagoneers and early Cherokees constantly show up in films despite their actual rarity on the road, in everything from other Netflix horror films like Bird Box to the inside art of Chris Stapleton albums.

28. Gilles Deleuze and Felix Guattari, *Anti-Oedipus: Capitalism and Schizophrenia* (New York: Penguin Books, 1977 [1972]), 277–278.

Chapter 5

"I Hope You Have a Big Trunk 'Cause I'm Putting My Bike in It"

Alternative Transportation as a Reinforcement of Capitalism

Bicycles don't fundamentally change anything, other than perhaps the fitness of the populace. We still would use them for commuting to work from our homes rather than question the very need to do work outside of the home in a capitalist manner that has led to the climate crisis. It is easier to envision the end of the gasoline-burning car than the end of daily commuting.

If we must rethink our lines of flight, as I argued in the last chapter, one option seems to be a focus on more human-powered vehicles. To return to Mackenzie Wark's problematization of the Anthropocene, apart from simply accepting that capitalism is inevitable and we must stick with it no matter the consequences, the only other framing of how to move forward is to envision going back in time to before we began to destroy the atmosphere.

> The alternative narrative imagines a kind of non-technical, holistic and spiritual alternative, often drawing its images from a pre-capitalist landscape. But as was already clear to Marx, this is *capitalist romance*, a story constructed within capitalism itself as one of the byproducts of its own momentum. It is a kind of capitalist realism in negative, where we all ride bamboo bicycles, but it rarely ventures beyond an ideological mirroring of capitalist realism.[1]

I want to focus on her line about bamboo bicycles, not just because it's funny, but to really interrogate this notion of returning to a simpler past through our mobilities. Having spent time both studying the geographies of cycling as well as plenty of time riding both recreationally and for commuting, I want to use this chapter to look at how alternatives to American automobility are presented.[2] I will first explore the concept of resistance to automobility in the cycling literature and then contrast that with its presentation in popular films. It is important to realize that cycling's framing as an opposition to a dominant ideology limits the analysis of the role of the bicycle in film. The reason cycling is represented as something weird characters do has more to do with the naturalization of the automobile than anything inherent to the bicycle. To demonstrate this I move away from cycling and to the Pixar film *Onward*, which argues for a return to a sense of wonder we once had, effectively constructing that capitalist romance of which Marx and Wark speak, all the while promoting American automobility in the process. What I am hoping to suggest with all of this is for us to question the underlying drive for movement that demands *either* a car or a bicycle. Too often, the hegemonic dominance of capitalism is overlooked in the fight to change our personal mobilities and a significant resistance is missed.

THE RIDER-BIKE ASSEMBLAGE: SYMBOLIC HUMAN-POWERED MOBILITIES

One argument for returning to simpler, nonfossil fuel burning mobilities like cycling is the idea that the benefits outweigh the risks of piloting a bike on automobile filled roads.[3] These quantitative studies often beg the question, at least implicitly, why wouldn't you ride a bike to work rather than your car? You live longer, feel better, look great, and you are reducing carbon emissions. Such studies evoke a political cartoon I show my students to facilitate a discussion on addressing climate change. One lecturing climate scientist is standing in front of a screen with a slide listing the benefits of switching to green technology: better jobs, clean air, healthy children, and so on. A member of the audience, however, stands up to ask, "But what if climate change is a hoax?" The point of course, is that even if it were a hoax, we would still live much better lives if we treated climate change like a real threat. These quantitative studies make the same argument; even if the societal and environmental issues with automobiles are not as bad as we think, riding a bike with make you happy and healthier nonetheless. What do we have to lose?

Other studies exist, of a more qualitative vein, that explore the political ramifications of choosing a bike over a car. Choosing a bicycle is not simply a logical choice for one's health but is also a personal action that can affect the larger community. Zack Furness contends that

> the bicycle is variously seen, and in many cases actively reconceptualized, as a source of self-empowerment and pleasure, a pedagogical machine, a vehicle for community building, a symbol of resistance against the automobile and oil industries, and a tool for technological, spatial, and cultural critique.[4]

The act of cycling is not simply a healthy activity that will make you live longer as the quantitative studies argue, but is simultaneously a way to resist the dominant ideology of American automobility.

Many of these foundational studies of cycling- or *vélo*-mobilities (as well as automobilities for that matter) were written around growing interest in Hubbert's decades-old peak oil hypothesis, in which the production of oil follows a bell curve in that it ramps up, peaks, and then drops off due to the finite nature of a resource like oil.[5] The rising costs of petroleum in the early 2000s suggested that we had or would soon exhaust our supply in the first decade of the twenty-first century.[6] Not only was climate change a direct result of our insistence on using inefficient personal automobiles, but diminishing oil supplies would raise the economic costs of maintaining this system. American automobility was posed to be both environmentally detrimental as well as much more of a class issue in terms of access. Notably, Kenneth Deffeyes, the scholar credited with the resurgence in Hubbert's hypothesis, was not calling for new modes of transportation so much as a new approach to energy use to maintain our existing way of life, including getting "over our nuclear phobia."[7] Peak oil did not happen as Deffeyes predicted, though a disputed recent study by BP argued that we did reach peak oil in 2019.[8]

Regardless of the moment at which peak oil happens, the ramifications of increased carbon emissions in the atmosphere are a real concern. Where someone like Deffeyes is looking at energy production and maintaining a way of life, others see a need to change our ways in order to combat overall environmental degradation. Again, the bicycle is a means to get to a cleaner, healthier Earth. Dave Horton examines how the bicycle is a mundane and practical material manifestation of the environmental movement:

> The bicycle is both symbolic, and iconic object of green discourse, and practical, an object in daily use, and furthermore one which lends distinctive form to the everyday lives of environmental activists, contributing to their green lifestyles and to the wider shaping of green culture.[9]

For Horton, the bicycle is a simple object that can do a lot. Symbolically and materially, the bicycle is a choice to slow down the increasing speed of modernity,[10] "open to all" in terms of gender and class,[11] and a performance of good community morals and healthy personal choices.[12] While I do not want to discount the value of the bicycle—a machine I hold dear—I see cycling as it is described by Horton and others as sliding into two of the four (ineffective) solutions Wark sees as being used to mitigate the effects of climate change. The bicycle allows one to consider one's own personal "carbon footprint" as well being "a romantic turn away from the modern, from technology, as if the rift [of climate change] is made whole when a privileged few shop at the farmer's market for artisanal cheese."[13]

My ultimate concern here is that in envisioning this environmental and socially just power immanent to the bicycle, we run the risk of both missing the social practices that imbue the bicycle with such virtue as well as promoting a "resistance" to the ideology of American automobility while nonetheless reproducing it. To be fair, Horton is not making the claim that all cyclists are environmental warriors or anything that simple. Rather he is suggesting that "it seems likely that the bicycle will continue to play a prominent part in individual and collective green projects in search of sustainability."[14] I would argue however that it is all too easy to view the bicycle as something opposed to American automobility rather than something in tension with it. My early thoughts on vélomobilities certainly mirrored the former and it was not until I moved to Northern Los Angeles County that I began to notice more complexity in the spatial and social differences inherent in mobilities. Where many of us stumble in our mobilities studies is that we assume all too easily that transportation modes have an essential quality, like the idea of cycling as resistance, even if we tell ourselves that we know better than to essentialize our objects of study.

The majority of cyclists in my own community are rather masculine as well as conservative in their political leanings. I have ridden and socialized with enough to know that they do not promote cycling in opposition to American automobility nor as a means toward a greener planet. This was clear when I first moved to the area and I, being the good greenie, rode my bike from my front door to the group ride meetup spots, while everyone else drove their bikes in the back of their trucks to get there. They didn't want to waste their legs getting to the ride. Cycling was about friendly competition. Conversations after these competitive rides also made it clear that these other cyclists did not see a need to reduce or eliminate inefficient automobiles. I also ran into many of these same cyclists at public meetings regarding improving local bicycle infrastructure, but they were less interested in being able to replace their cars so much as have a safe place to exercise and play on the weekends when they weren't driving to work. Driving to work also provided the means to buy that newer, lighter machine to add to one's stable of bikes. While they would get angry at a motorist driving too close as they rode or cutting them off, these cyclists never framed it as cars are the problem, but rather, bad drivers are.

As far as my latter concern of resistance, too often we romantically invoke Foucault (odd considering his work, I know), by citing his axiom "where there is power, there is resistance" while never interrogating the second half of the statement, "and yet, or rather consequently, this resistance is never in a position of exteriority in relation to power."[15] Foucault contends that these points of resistance exist within a large swath of power relations which mean there is no singular "soul of revolt."[16] Žižek takes this further though, arguing that power and resistance are even more intertwined than Foucault conceived: "Power is always-already its own transgression, if it is to function, it has to rely on a kind of obscene supplement."[17] In other words, the resistance to power of which Foucault speaks is not only internal to the relations of power, but it is also not necessarily a *challenge* to that power. Using this Žižekian read of power/resistance to understand cycling and American automobility, the resistance in the act of riding a bicycle has already been accounted for by this dominant ideology as a pressure valve of sorts. The perceived resistance in bicycling in order to be one less car on the road can in fact be a symbolic and personal victory, but it does not ultimately prevent the dominant ideology from persevering. If anything, one less car makes for less traffic and a more enjoyable commute! If I am a good subject of automobility, the only thing that makes my commute to work problematic is that cyclist who insists on sharing the road.

The dominant system of American automobility is obviously an unsustainable system to anyone who studies it. If we continue with Žižek we must remember that any ideology uses fantasy *to take its own failure into account in advance*.[18] That is, any inconsistency within an ideology is addressed through some sort of fantasy to cast blame not on the ideology but upon some other cause. In the case of American automobility, the inconsistencies inherent with maintaining a system of personal automobiles in a dense society are always the fault of other drivers, corrupt or ineffective politicians, greedy car companies, or anything other than the system itself. The symbolic Other Driver, for example, is always to blame for traffic (because they don't drive aggressively enough, or too aggressively, etc.) rather than the idea that thousands of driver-car assemblages simply cannot exist in the same place at the same time. The very deviance of the cyclist is also a useful tool in maintaining American automobility. Take the government funding that goes into bike lanes and separated bike paths; American automobility is not sustainable, but by simultaneously building and underfunding bicycle infrastructure one can point to how empty these bike lanes are. Only the occasional outlier uses such transportation modalities, thus reinforcing the dominance and natural state of driving a personal car or truck. When cycling activists advocate for an increased share of transportation funds, local governments can point to the absence of cyclists on the infrastructure that already exists. We even see segregated bike paths installed to appease cyclists, removed to appease motorists, and then reinstalled yet again.[19]

Rather than a sustained look at what resistance to American automobility would really look like, qualitative scholars of vélomobilities continue to imbue the bicycle with mystical powers of bringing people together to change society. The bicycle is not unique here; the driver-car assemblage I have been discussing is not due to a mystical power transcendentally inherent to the modern automobile either. My point is that a vehicle, whether human-, electricity-, or fossil-fuel-powered, is symbolically fluid depending on what the human component of the assemblage adds to it. When I ride my bike, either to commute to work or for recreation and exercise, I fully admit that I do so in a way that fits with Horton's analysis. I chose to start riding a bike to make a political statement as well as to see if I could move from one part of the city to another without dying from either exhaustion or getting hit by a car. Successfully riding the few miles that first time was exhilarating and I was hooked. I switched jobs shortly after that first utilitarian bicycle trip and now had a commute of about fifteen miles by car using freeways that I shrunk to nine miles cutting across a railyard and a small bridge. I still remember a motorcycle cop next to me at a red light during one of my morning commutes. He asked about the route and distance and then called me "hardcore" as the light changed and he rode off. Even "the man" was acknowledging my commitment to cycling and resistance! But I soon realized that my personal choice was simply that, personal. This rider-bike assemblage that I had crafted facilitated a new mobility for me, but it is important to acknowledge that it was personal to me within the dominant system of American automobility. My "hardcore" assemblage led to a physical and ethical becoming, but even as a perceived resistance it did nothing to challenge the existing order of things. I am not discounting the value of reducing even a small amount of carbon emissions or experiencing a place from the vantage point of a bicycle saddle. I simply question the idea that we can replace one vehicle with another to finally achieve a *good* form of American automobility.

The logical question is then, well, how exactly should we resist automobility? If this system is so bad, we ought to get rid of it, but how? Böhm et al. argue that we must "expose the inconsistencies, contradictions and antagonisms of the present regime of automobility."[20] The problem with this approach is that these inconsistencies have always been there, but fantasy allows them to be pushed onto something else. If simply pointing out that an action was not in one's best interest could change life for the better, society would be a much different place. Do racists hate others simply because they don't understand that biologically speaking racial classification is not valid? Of course not.

Žižek is one of the few intellectuals to suggest a way forward despite the ineffectiveness of resistance. The trick is to "subvert the ruling ideology by taking it more seriously than it is ready to take itself."[21] In other words, if we conform to ideology without any compromise, we find that the ideology is not tenable. A major component of Žižek's theory of ideology is the necessary distance between our actions and that dominant code. Using Stanly Kubrick's

Full Metal Jacket as an example, he argues that Vincent D'Onofrio's character, Private Pyle, is a failure of military subjugation precisely because he tries to conform too hard and snaps. To fully embrace the role of a soldier in a place and time like the Vietnam War is not something a human can maintain. Matthew Modine's character, however, a detached soldier who keeps a distance from the ideology behind the military presence in Vietnam, becomes a subject of the "military big Other" when he shoots a wounded Viet Cong woman—"he is the fully constituted military subject."[22]

What would an overconforming to the ideology of American automobility looks like? Furness describes a Critical Mass bicycle event in the opening chapter of *One Less Car*:

> Originating in San Francisco in 1992, Critical Mass was conceived as a group bike ride and leaderless celebration that ultimately grew in both size and popularity as a response to the continued marginalization of bicycling and non-motorized transportation in modern cities. Each month cyclists taking part in this "organized coincidence" try to fill the streets with riders to demonstrate their collective solidarity and send a message to the public: "We are not blocking traffic; we *are* traffic!"[23]

The problem with this resistance to a dominant, car-based automobility, however, is that, as even Furness points out, it led to the arrest of 264 cyclists, but no change in transportation policy. This is not to suggest that a single protest of any type is even capable of changing something as big as our dependence on the automobile. Police, however, argued that the "massive disruptions" of the Critical Mass event blocked "people trying to get to the hospital."[24] Regardless of the validity of the New York Police Department's justification, such an action reinforces the idea of cyclists as being outsiders to the system of automobility. Böhm et al. have pointed to the normalization of driving through the othering of alternative mobilities. "Cyclists, for example, are routinely rendered as deviant, both in planning processes which assume their non-existence, and where the car driver is manifestly the 'normal subject,' and in more active moral panics such as the one about 'Lycra louts.'"[25] The deviance given to nonautomobile travel within the dominant ideology of American automobility limits the power within a form of resistance like Critical Mass. One expects the cyclists to protest these kinds of things like rights to the road because they are simply different from us, the good American driver.

Rather than an overt refutation, Žižek suggests that the opposite form of protest is needed. The *perfect* American motorist can ultimately show the failure of the system. What would happen if these cyclists actually drove hundreds of cars into the streets? The effect would be the same—drawing attention to the problem of individual cars clogging up streets so that they cannot be used for other purposes—but it would be done by fellow motorists. By these hundreds of protestors doing exactly what they are "supposed" to do in independently driving hundreds of cars into the same intersection, they would shut down the system by conforming to it.

Perhaps the best film to display the absurdities of American automobility is Mick Jackson's 1991 film, *L.A. Story*. The automobile is a dominant component of life even if it becomes clear that it is making life more difficult and even causing Angelinos to imagine repeated heart attacks. Early in the film, Harris K. Telemacher (Steve Martin), a television weatherman, gets in his car to drive to work. The camera is tight on Harris, but we can hear the cacophony of car horns honking. He turns on the radio in his car to the traffic report which announces major gridlock and the camera cuts to a dozen cars tightly crammed on a street. We see one driver completing her cross-stitching as she sits in the traffic jam that has no indication of ending anytime soon. It turns out the traffic comes right up to Harris' house, so without missing a beat, he drives his car up onto the sidewalk and past the stopped cars. The camera cuts to a shot of Harris using both hands to buckle his seat belt as he drives—safety first. His eyes are on the belt latch and not the road. Harris turns onto a driveway. A woman, presumably the owner, stands on her balcony in a pink bathrobe. She waves warmly to Harris as he speeds up her driveway and through her backyard. He returns the wave, indicating that this isn't the first time he has taken this route to work. The journey gets more and more bizarre, with Harris fixing his hair as he drives into the L.A. River, up a steep gravel road that appears to be well outside of the city and then down an outdoor set of concrete stairs. He finishes the drive through a series of lawn sprinklers that leave his car looking brand new. He pulls into his parking spot at the television station right on time.

Harris Telemacher is the perfect subject of the ideology of American automobility. We can tell because no one is upset by his driving—in fact, the people he drives past are happy to see him. Harris has fully committed to the individualism and freedom promised to subjects of the dominant ideology, so much so that he uses the car for even the simplest of tasks, like driving his car two houses down just to visit his friend. As a character, Harris is both committed to the ideology, but at the same time makes aside comments to let the viewer know that this whole thing is supposed to be a comedy. We enjoy the film because we are outside of its world; at least, we think we are outside, maintaining some sort of ideological distance (Figure 5.1).

Not only is the automobile an important component to *L.A. Story*, but the freeways of Los Angeles and other infrastructure play a key role in the story. One night, Harris' car unexpectedly breaks down right in front of an electronic freeway sign that reports on current traffic conditions. As he examines

Figure 5.1 Harris Telemacher (Steve Martin) on His Normal Commute to Work in *L.A. Story*. *Source*: *L.A. Story*, dir. Mick Jackson (1991)

the car's engine, the sign begins to communicate with him by displaying messages directly to him, like "R U OK?" Harris goes along with the sign, at first thinking someone is playing a prank on him, but he is even willing to hug the signpost simply because it politely asks for one. "I C PEOPLE N TROUBLE & I STOP THEM," the sign explains to Harris and then gives him a riddle to figure out what he needs to do to help himself. Just as suddenly as the car stopped, it starts again, presumably the result of this magic signpost. Throughout the rest of the movie the signpost will communicate with Harris, giving him advice to live a more authentic life. Harris explains in a voiceover narration that he cannot explain how this event with the signpost happened and chalks it up to an example of magic. For magic to be channeled through a freeway signpost rather than the art galleries or restaurants Harris visits throughout the film is a powerful reminder of just how natural the automotive landscapes of Los Angeles have become. The film speaks to the wonder to be found all around us, but real magic in this film comes from random weather events and automobility, thus naturalizing the place of the car.

The absurdity of *L.A. Story* works precisely because it reveals what would happen if we fully commit as subjects of American automobility. The humor comes from the fact that we know that this behavior should never happen and yet it is exactly what the dominant ideology seemingly demands. *L.A. Story* presents what an overconformity to ideology looks like. On screen it is hilarious, in reality the system would collapse.

When looking at how cyclists are presented in popular films, the opposite takes place. A cyclist is presented as being deviant just as Böhm et al. suggest. Furness acknowledges this in *One Less Car* in a chapter dedicated to exploring media depictions of the bicyclist. "In the few cases where adult bicyclists are featured in U.S. entertainment media, they are generally portrayed as being far outside the mainstream; most are depicted as childish men, eccentrics, sexually off characters, geeks, and/or financial failures."[26] Furness thoroughly covers the handful of popular films that involve cycling like Tim Burton's classic *Pee-Wee's Big Adventure* and Judd Apatow's *The 40-Year-Old Virgin*. For Furness, the problem with representations of cycling men (always men) is that they are presented as losers or freaks, even if the audience roots for them. In *Pee-Wee's Big Adventure*, Pee-Wee (Paul Reubens) is an eccentric person with the sense of wonder and play of a boy in a man's body whose most prized possession is his bicycle, a red and white machine with accoutrements like handlebar streamers and a stylish chain guard. The plot of the film is pretty straight forward; Pee-Wee's wealthy neighbor Francis (Mark Holton) has the bike stolen, but then panics and tries to get rid of it. The theft leads Pee-Wee on a trip through the United States to recover the bicycle. "Within the exaggerated, campy world that Pee-Wee inhabits, his bicycle obsession seems normal given the fact that he plays with toys, shops at magic stores, wears fuzzy bunny slippers, and lives in a house resembling a small carnival."[27] Pee-Wee is lovable and the audience cannot help but root for him on his quest, but as Furness points out, the audience does not want to be just like him and emulate his mobility choices.[28] Further, while Pee-Wee is quite competent on his bicycle, other, more masculine or adult forms of transportation are beyond his reach. In one of my favorite scenes in the film, Pee-Wee has befriended a group of bikers in a bar who give him one of their motorcycles, as well as a leather vest, to help him on his quest. Pee-Wee thanks the enthusiastic bikers who cheer him on as he starts the bike. He rides off and only gets about 20 feet before he runs into the bar's sign and is flung off the motorcycle.

Furness is clearly a fan of *Pee-Wee's Big Adventure* and is not suggesting that it is offensive to cyclists, but rather it is one of many films that promote and image of the bicycle as the transportation mode of the outcast or loser. Where someone might be inspired and excited to drive a car after seeing a film like *Ford v. Ferrari*, it is doubtful that *Pee-Wee's Big Adventure* encouraged any viewers to ditch their cars and start riding to work. What Furness misses though, is that from an ideological standpoint, the film is questioning masculinity rather than directly troubling American transportation. Yes, the man-child Pee-Wee has a bicycle rather than a car, but nothing in the film gets at the naturalization I have been arguing ideology requires to fully capture subjects. If anything, as Furness acknowledges, the final scene in the film has Pee-Wee attending the premiere of a major Hollywood film about his life and adventures. James Brolin and Morgan Fairchild play Pee-Wee and his love interest, respectively, and the silly red bicycle has been replaced by a sexy red motorcycle. Pee-Wee himself has a cameo in the film, but his voice has been dubbed with a much deeper one. Clearly, Reubens and Burton understand the hyper-masculine absurdity of Hollywood action films and are mocking them rather than reinforcing the idea of the dominance of the automobile.

The 40-Year-Old Virgin, while also an absurd comedy, is more true to life in its depiction of mobilities and masculinity. Steve Carrell plays the titular character Andy Stitzner, who not only has yet to have sex, but also collects action figures, has a lowly service job at an electronics store, and rides a bicycle. And as if riding a bicycle wasn't bad enough, Andy does so in a way that could not be conceived of as even slightly cool. The bike has fenders, upright handlebars, and a large red basket attached to the back. Andy also rides in khaki pants that he folds and tucks into his sock and uses hand signals to indicate turns to further remove him from the image of an athlete or cool bike messenger. Later in the film, his original bicycle has been replaced by a new dual suspension Trek mountain bike with a much sleeker matching helmet. Notably, shots of Andy riding the Trek still have him wearing normal clothes, but his pants are less nerdy and not tucked into his sock. *The 40-Year-Old Virgin* certainly works to make cycling a deviant form of mobility as well as connect it to unusual sexualities. And yet, what is most interesting is that getting into a heterosexual relationship and finally having sex does not change appear to change his mobilities choices.

> Indeed, the only surprise of the film is that Stitzner does not ride off in a brand-new automobile after having sex for the first time—a move that would have secured his status as an authentic male in the eyes of American filmgoers. He remarkably manages to become a real man without having to give up his bike.[29]

A big part of Andy's character is to stay true to who he is despite pressure to change. He resists selling his collectable action figures and resists the pressure his friends put on him to have sex despite their own miserable relationships. The very fact that he doesn't drive off in a car at the end of the film speaks to a larger theme of existential authenticity over traditional masculinity. Bicycling in *The 40-Year-Old Virgin* is absolutely played for laughs as something that successful grown men do not do, but the film does not overtly reinforce the natural "common sense" of American automobility in a Gramscian sense. The real power of popular films has not so much made cycling deviant as it has given the automobile a natural choice for travelling the landscape, no matter the context. For the bicycle to be presented as something else, filmmakers could push for a "shift of accentuation" to alter the deviance assigned to it, but I think a larger critique of common sense in moving via car is needed first.[30]

RECLAIMING MAGIC IN A VAN: PIXAR'S *ONWARD*

For Gramsci, common sense is "a cultural battle to transform the popular 'mentality' and to diffuse the philosophical innovations which will demonstrate themselves to be 'historically true' to the extent that they become concretely—i.e. historically and socially—universal."[31] That universality is the real key behind this kind of ideological power in that it makes what should be socially contingent into the natural order of things. The deviance of the bicycle as discussed above is only invoked because the act of driving a car has had decades of work to make it the natural way to move about city streets, dirt roads, and so on. Bicycles are so rarely used in films precisely because the default and "historically true" mobility is the automobile.

Pixar Animation Studio's *Onward* tells the tale of two elf brothers on a quest to resurrect their dead father for just one day. The film opens with narration explaining that long ago, the world was full of wonder, adventure, and most important, magic. While mastering magical spells was difficult, inventions like the light bulb and modern kitchen appliances eliminated the need for it, leading to a mythical world of elves and centaurs that looks rather suburban. The home of brothers Ian (voiced by Tom Holland) and Barley (voiced by Chris Pratt) might be shaped like a mushroom, but it nonetheless has a manicured lawn, hedges, and a typical car parked on the driveway. The problem with this version of modernity is that life has become too safe and boring.

Both brothers are outcasts of a sort. Ian is too shy to even try to fit in with other kids and despite the film beginning on his sixteenth birthday he is too nervous to learn to drive. Barley, the older brother, is a nerd and a screw-up, albeit in a lovable way, obsessed with "Quests of Yore," a Dungeons and Dragons type role playing game. Barley also drives a purple van he has named Guinevere and has painted a winged horse along the side. The Quests of Yore game ties into Barley's overall passion for all things ancient and magical.

Early in the film, Ian takes and fails his driving test. It isn't a prolonged scene; he simply panics when asked to merge onto a busy freeway and shouts "I'm not ready!" This is the first of several masculine and adult goals that he has set for himself on his birthday and promptly fails, including to "be more like Dad." The fact that the driving test is focused on is a testament to its natural positioning as a rite of passage to adulthood, even in this mythical version of Earth.

Their father died of an illness before Ian was born. Before he passed, their dad left a gift of a wizard's staff for both boys to open once they were older than sixteen, as well as a spell that could resurrect him for one day. Barley cannot make the spell work, but Ian clearly has a gift for magic. He recites the spell and it begins to make his father appear. Something goes wrong with the staff's gem though, and the spell stops working, leaving just their father's animated legs clad in slacks and loafers. The two brothers then go on a quest to find another gem to complete the spell. The quest introduces the brothers to a cast of mythical creatures who have allowed modern conveniences to tame them, like the Manticore, now known as Corey (Octavia Spencer) who manages a theme restaurant. Her responsibilities for the mundane administrative tasks of the restaurant have robbed her of the passion she once had for adventure. The brothers help restore her sense of wonder, as well as bond with each other and understand more of what it means to be a family. But what is telling is the fact that despite the film's other theme of returning to a magical past of adventure and of eschewing the modern conveniences that have made life dull; the van is one modern convenience that is *necessary* for adventure.

On the quest, Ian overcomes his fear of driving and pilots Guinevere along mountain roads to escape the police. It is clearly a moment in which he has changed for the better and is on his way to becoming a man. When it is obvious that they cannot get away from the cops, Barley sacrifices her by placing a rock onto Guinevere's gas pedal and sending her into a mountain side to knock boulders onto the road. As the van drives toward the mountain, a tire pops and the sound of rubber knocking against the van sounds like a galloping horse. The scene is dramatic and emotional as the van is crushed.

The fact that the van is made an important and even heroic part of the film speaks volumes to the naturalized idea of automobility. When thinking of quests and magic, my first thought is a film like *The Fellowship of the Ring*, which if anything, is the ultimate hiking movie. Once the van is destroyed, the boys and their father's legs do go on a hike for a bit, but the film is pretty much in the third act at that point. The issue here is that for all of the magic in the film, the van is treated as a natural component of a universal rite of passage, whereas every other part of modern society gets in the way of the quest. Cycling will always be different from driving in films when they are produced within this ideology. The latter mobility is a common sense, natural way to travel across practically any space, whereas the former is something fun enough for kids or athletes, but not what you choose when you have to cross through ecosystems.

At the end of *Onward*, the characters have all changed for the better as one would expect from a Pixar film. Barley, however, is able to return to his pre-quest self with a new van that has another winged horse painted on its side, though this time with Ian and Barley painted riding it. The van is important after all; as it must be there should they ever have another quest on which to embark.

RESISTING THE MAGIC OF THE CAR

Films like *L.A. Story* and *Onward* highlight the, absurd as it sounds, magic qualities we have given the automobile. When we consider American automobility as an ideology, however, we should not be surprised that we have gotten to this point. As Stuart Hall contends with his Gramscian read of what ideology is and does, "The point at which we lose sight of the fact that [common] sense is a production of our systems of representation is the point at which we fall, not into Nature but into the naturalistic illusion: the height (or depth) of ideology."[32] As a society we have long since forgotten the all too human systems of representation that have gone into normalizing the automobile. Its role in both everyday life and extraordinary events is unquestioned and will continue to be treated as such until we perform an ideological critique as well as a Foucauldian genealogy of the automobile.[33]

As far as a resistance to this dominant ideology, the bicycle alone will not be enough to address the climate crisis. While we ought to take a bike whenever possible, or better yet, question the apparent need for movement and its connection to consumption and capitalism, a more systemic fix is needed. Again and again, I return to a Žižekian conception of power and resistance, one that accepts that power has already accounted for its weak points and inconsistencies. What happens if we focus more on leaning into automobility like Harris Telemacher?

COVID-19 has presented me with another example of the complexity of political resistance. In Los Angeles County, cases of the disease have exploded throughout the pandemic leading to waves of tight restrictions on businesses temporarily closing or suspending some services. The small, rural community in which I live is a rather conservative, if not reactionary place. Even after the 2020 election, Trump flags still fly from a number of flagpoles around town. As such, many people, business owners and otherwise, have complained about the government's response to the pandemic, targeting California governor Gavin Newsom and Los Angeles Mayor Eric Garcetti as fascists who are ruining our home. I have taken to using what little interactions I have with people these days to feel out the conservative stance on dealing with the pandemic. I have yet to run into anyone locally who thinks it is a hoax and many of the people I have spoken with understanding that it is a severe disease (although many assume they don't fit the profile of those who are most at risk of complications despite evidence to the contrary). In a lengthy discussion I had at the local feed store, one of the employees complained that while he and his bosses want to keep their customers safe, they do business in a conservative place and cannot ensure that every customer wears a mask (the employee himself was not wearing a mask at the time). He continued on about government regulations, but then immediately shifted to complaining about the lack of government support to address the disease. He was asking for a stronger government in terms of financial assistance to help keep average people able to stay home and to help businesses pay for the changes they need to make to comply with regulations. Effectively, in this land of conservatives who dislike anything remotely resembling socialism, the COVID-19 pandemic has made many of them look for a better government response. The methods to combat COVID-19 here in California have in many ways felt like overconforming to the conservative neoliberal system that the United States has been pushing since the 1980s. It is a mix of prioritizing some private businesses while offering no support to the labor needed to perform essential services or to stay home to limit the virus' spread. While the conservatives to whom I have been speaking still identify with their conservative political leanings and use Newsom or Garcetti to "stitch up the inconsistencies" of this ideology, this overliteral conformity to neoliberalism is showing just how untenable it is for the majority of us.[34] If overconforming to neoliberal privatization and austerity can make a Trump supporter ask for socialism, what would it take to get us to finally leave our cars parked?

NOTES

1. McKenzie Wark, *Molecular Red: Theory for the Anthropocene* (London: Verso: 2016), xxi, her emphasis.
2. Michael W. Pesses, "Automobility, Vélomobility, American Mobility: An Exploration of the Bicycle Tour," *Mobilities* 5, no. 1 (2010): 1–24.
3. David Rojas-Rueda, Audrey de Nazelle, Marko Tainio, Mark J. Nieuwenhuijsen, "The Health Risks and Benefits of Cycling in Urban Environments Compared with Car Use: Health Impact Assessment Study," *BMJ* 343, no. 7819 (2011): 1–8; Jeroen Johan de Hartog, Hanna Boogaard, Hans Nijland, and Gerard Hoek, "Do the Health Benefits of Cycling Outweigh the Risks?" *Environmental Health Perspectives* 118, no. 8 (2010): 1109–1116.
4. Zack Furness, *One Less Car: Bicycling and the Politics of Automobility* (Philadelphia: Temple University Press, 2010), 9.
5. Kenneth S. Deffeyes, *Hubbert's Peak: The Impending World Oil Shortage* (Princeton, NJ: Princeton University Press, 2002).
6. Deffeyes, *Hubbert's Peak*, 1.
7. Deffeyes, *Hubbert's Peak*, 189.
8. David Blackmon, "So, We're Back to 'Peak Oil' Again," *Forbes* Sept. 23, 2020, https://www.forbes.com/sites/davidblackmon/2020/09/23/so-were-back-to-peak-oil-again/?sh=9d8488e2c9a6; Notably demand for oil is significantly down as I write this thanks to stay at home orders for much of the United States.
9. Dave Horton, "Environmentalism and the Bicycle," *Environmental Politics* 15, no. 1 (2006): 42.

10. Paul Virilio, *Speed and Politics: An Essay on Dromology* (South Pasadena: Semiotext(e), 2006 [1977]).
11. Horton, "Environmentalism and the Bicycle," 45.
12. Horton, "Environmentalism and the Bicycle," 46.
13. Wark, *Molecular*, xv.
14. Horton, "Environmentalism and the Bicycle," 55.
15. Michel Foucault, *The History of Sexuality, Volume 1: An Introduction* (New York: Vintage Books, 1990 [1978]), 95.
16. Foucault, *The History of Sexuality*, 96.
17. Slavoj Žižek, *The Plague of Fantasies* (London: Verso 2008 [1997]), 35.
18. Slavoj Žižek, *The Sublime Object of Ideology* (London: Verso 2008 [1989]), 142, his emphasis.
19. Jake Offenhartz, "Sometimes the City has to Destroy a Bike Lane Before it Can Reinstall a Bike Lane in the Same Place," *Gothamist,* November 4, 2019, https://gothamist.com/news/sometimes-city-has-destroy-bike-lane-it-can-reinstall-bike-lane-same-place; Janette Sadik-Khan, "The Bike Wars Are Over, and the Bikes Won," *New York*, March 8, 2016, https://nymag.com/intelligencer/2016/03/bike-wars-are-over-and-the-bikes-won.html.
20. Steffen Böhm, Campbell Jones, Chris Land, and Mat Paterson, "Introduction: Impossibilities of Automobilities," *The Sociological Review* 54, no. s1 (2006): 9.
21. Žižek, *The Plague of Fantasies*, 99.
22. Žižek, *The Plague of Fantasies*, 27.
23. Furness, *One Less Car*, 1–2, his emphasis.
24. Johnathan Wald, "264 arrested in NYC bicycle protest," *CNN*, August 28, 2004, https://www.cnn.com/2004/ALLPOLITICS/08/28/rnc.bike.protest/.
25. Böhm et al., "Impossibilities of Automobilities," 8.
26. Furness, *One Less Car*, 109.
27. Furness, *One Less Car*, 110.
28. Furness, *One Less Car*, 111.
29. Furness, *One Less Car*, 111.
30. Stuart Hall, "Signification, Representation, Ideology: Althusser and the Post-Structuralist Debates," *Critical Studies in Mass Communication* 2, no. 2 (June 1985): 113.
31. Antonio Gramsci, *Selections from the Prison Notebooks*, edited and translated by Quintin Hoare and Geoffrey Nowell Smith (New York: International Publishers, 1971), 348.
32. Stuart Hall, "Signification," 105.
33. For a start to such a project, see Michael W. Pesses, *Crawl Space: Driving over the Anthropocene in a Jeep*, PhD diss. (Claremont Graduate University, 2020).
34. Žižek, *Sublime Object*, 49.

Conclusion

Early on in my graduate school days, an influential professor passed along Christopher Salter's essay "The Cowboy and the City: Urban Affection for Wilderness." I expressed interest in the piece simply because it was about the Marlboro Man, that frontier-obsessed marketing campaign by Marlboro cigarettes, but Salter's thesis wound up becoming a break in my own intellectual development. His argument revolves primarily around the plight of urban residents in twentieth-century America and contends that most urbanites (at the time of the essay anyway) are migrants to the city from a simpler rural upbringing so therefore view the city as outsiders. This, explains Salter, is why Marlboro's cowboy campaign was so successful. Marlboro cigarettes initially played up a European sophistication in their marketing, but in switching to the cowboy affect they became the most globally popular cigarette by 1973.[1] In all of the advertisements using the Marlboro Man theme, the cowboys depicted maintain a consistency: "Here is our composite: a solitary man, strong and confident against a backdrop of majestic virgin nature, working his animals with a masculine grace, unassisted by any but the most rudimentary technology."[2] Advertisements would often have the tagline "Come to Marlboro Country" to produce a sense of place to go along with the cowboy figure. The obvious conclusion regarding Marlboro's success is a desire to be just like him. If I buy a pack of Marlboros it is because I am a cowboy at heart even if I am stuck living in a city. But, Salter contends, our desire to live like the Marlboro Man is a myth. "We covet the image. We avoid the reality."[3] It really would not be that hard to live such a life, and yet every year more and more Americans move to cities and fully embrace the technological and social conveniences that the Marlboro Man eschews.

It is all too easy to pull a Marlboro ad from the 1970s or 1980s and critique its sense of place. We can point to the overall colonial motif in that "Marlboro

Country" exists on once indigenous lands. The Marlboro Man, despite being portrayed by multiple actors, is always a cisgender white man, thus producing societal rules as to who belongs in the landscape and who does not. And of course, while nature is often a backdrop, as Salter explains, it is most definitely a nature sculpted by humanity despite its pristine appearance:

> The scale of the setting is distinctive. It is most often monumental, as suggested by high jagged mountains, open, snow swept plains, and also powerful animals being tended. Here, then, the lone individual is nested not only in untouched nature, but in majestic, even stunning, nature. The setting is powerful, yet it is just half of the image. The cowboy himself creates the rest.[4]

We can direct our critical gaze to all of these aspects of the Marlboro Man, and yet, what do we have to show for it? In other words, does the acknowledgment that the masculinity and nature presented by the Marlboro Man is problematic actually do anything to stop the ad campaign's ideological hold? What we need to be doing here is to work to get beyond a superficial list of grievances with popular culture and dig into the ideological work being done to those consuming it.

Salter's concluding points are that there is a curse to the Marlboro Man in that in longing for pristine landscapes, urbanites fail to see the beauty of the city. There is an aesthetics within the built landscape that is no less valuable than Marlboro Country. When first reading the essay, I was less interested in the beauty of the city and more about the apparent contradiction in urbanites desiring the cowboy lifestyle. What we are doing when we reach for that Marlboro, is we grab a little piece of the frontier experience, but we have no intention of actually living like him. For all its perceived simplicity, living the life of a cowboy, at least as depicted in these ads, is hard. Modern conveniences make life so much easier. So rather than take the steps to live in Marlboro Country, we simply buy a pack of cigarettes to signal that we are cowboys at heart.

The value of something like the Marlboro Man, ideologically speaking, is tremendous. If we agree with Salter's thesis that these are urban migrants doing the smoking, we can position them as labor within a capitalist system of production. Marx's copious analysis has shown that labor within such a system is limited in its power in that "the possessor of labor-power, instead of being able to sell commodities in which [their] labor has been objectified, must rather be compelled to offer for sale as a commodity that very labor-power which exists only in [their] living body."[5] The system is stacked against labor, and yet, it persists. Ideology, lurking within state apparatuses, as Althusser argues, is what allows for the reproduction of this system in maintaining good subjects.[6] We can extend Althusser's concept past schools and law enforcement and use media and popular culture to explore how consumption affects the ideological hold. This is what a big part of Stuart Hall's project was:

> A critical question in developed liberal democracies is precisely how ideology is reproduced in the so-called *private* institutions of civil society—the theater of consent—apparently outside of the direct sphere of play of the State itself. If everything is, more or less, under the supervision of the State, it is quite easy to see why the only ideology that gets reproduced is the dominant one. But the far more pertinent, but difficult, question is how society *allows* the relative freedom of civil institutions to operate in the ideological field, day after day, without direction of compulsion by the State; and why the consequence of that "free play" of civil society, through a very complex reproductive process, nevertheless consistently reconstitutes ideology as a "structure in dominance."[7]

Studying a repressive, authoritarian state is easy, but how can we explain the reproduction of a system that exploits labor in a free society? For Hall the answer lies in Gramsci's hegemony, consent, and common sense. Ideology is reproduced in those "so-called private institutions" to have labor go along with the dominant ideas. Private interests will encode their messages with, typically, those dominant ideas, but it is also up to the consumer to decode those messages in a way that either accepts or critiques what they receive.[8] The limitations of such an analysis, however, become clear with the Marlboro Man. This individualistic, rugged cowboy is the antithesis of the capitalist worker, so it cannot be as simple as deciding to accept or challenge the message from a dominant group. To grab a pack of Marlboros because of a cowboy ad campaign is to toy around with the idea of being a rugged individual, but the key here is Salter's argument that "we avoid the reality."[9] There is nothing at odds between a labor force prizing individualism if the individuals have no intention of working to get rid of their ideological chains.[10] Žižek's attention to psychoanalysis allows us to grapple with the seemingly contradictory images and ideas that flood modern life. Fantasy, like the idea of our independent and competent Marlboro smoking cowboy, *is a means for an ideology to take its own failure into account in advance.*[11] The failure of capitalism here being the very exploitation of workers. A fantasy is a way to let subjects deal with an inconsistency, but to never force an actual fix. It is a pressure valve that maintains the system precisely because the system is still kept hidden. In taking a long drag off a Marlboro cigarette, I can fantasize about living the macho, rugged life in the brief moment I have, and then go right back inside to my warehouse job that exploits me. Following this reasoning, our desire for Marlboro Country is Lacanian, it is meant to remain symbolic rather than become a new reality.

Ecomobilities has not been about cigarettes though. What I have tried to argue for here is a reconfiguring of how we connect the automobile to both physical, natural places and our social construction of nature itself. As we acknowledge and grapple with the Anthropocene, we must absolutely seek out the problems that have led to a rapidly changing climate, and of course, a distinctly American reliance on the personal car or truck has been a major contributor to both local and global environmental health. The problem, however, is that all too often I have read about the environmental problems of the personal car, in addition to the health and safety concerns, as if that were enough to get us to rethink our mobilities. The automobile, in the American sense I have been using throughout this book, is ideological. Rather than a simple tool, the assemblage produced by humans, machines, capitalism, and nature facilitates a mobilities that goes beyond simply moving from point A to point B. The car is something with which we assemble to do things that we could not do without the machine. Or is this really the case? The power of ideology lies in its ability to shape desire even if we recognize its absurdity.

Just like the symbolic desire of Marlboro Country, American automobiles, trucks, and SUVs mainly represent an *objet petite a* that we have seen in the films discussed above.[12] We buy a Jeep Rubicon or Ford F-150 to grab that little piece of an adventurous life, but we have no real intention of taking the steps to live such a life. To drive a truck off road is to potentially damage it, get it stuck, get horribly lost, and so on. The sheer cost of today's full size pick-up trucks almost demands they be kept on the pavement to protect them. And yet, the truck *could* conquer a mountain or haul heavy equipment if the need arose. I would suggest that there is an extra absurdist fantasy to all of this, although maybe no more absurd than the Marlboro Man. A truck or SUV is not simply about driving over rocks to get out into nature, but also a tool necessary in dealing with a zombie apocalypse or Kaiju assault. This is made doubly complicated when we realize that these horror and sci-fi films are using those monsters to represent the anxiety producing environmental degradation that, at least in part, those trucks and SUVs have contributed to. The films we watch are certainly not wholly responsible for American automobility, but they certainly add to its reproduction and to the endless feedback loop of American automobilities and climate change.

There is an ideology at work with American automobility that precludes the possibility of using rational argument to escape it. To list the environmental, social, and public health problems with the dominant system of personal transportation that exists in the United States and other countries is to think in terms of homogenous space as Bergson critiqued.[13] It renders the actual experience of driving a car as something inert, dead, still. American automobility is a much more complex amalgamation of machines, humans, space, time, nature, art, capitalism, and ideas that cannot be stopped by facts.

The question we must first ask is whether we can, should, or must imagine the end of the personal car. The efficacy of our question will lie in our ability to realize the Gramscian common sense that has been produced and accepted when it comes to our own mobilities. We have been interpellated by the ideology of American automobility to the point that we can only see one way to travel across an ecosystem. I have long struggled balancing the knowledge of the ills of the fossil fuel burning vehicle, living in suburban environments that demand such mobility, and well, thinking some of those trucks are pretty cool. What is interesting is the shock of the COVID-19 pandemic has led so many of us to question our mobilities. Do we need to commute? Will our lifestyles work in a new epoch? What happens if we slow everything down, and do so together, not leaving the most vulnerable behind? It might be too late to avoid the magic number of 1.5°C, but it will never be too late to fight the inequalities that will no doubt make the Anthropocene much harder for some.

In Tony Scott's *True Romance*, Clarence (Christian Slater), when discussing to where he and his wife Alabama (Patricia Arquette) will run away says, "I've been in America all my life. I've always wanted to see what TV in other countries looks like." Popular media is the lens through which we Americans engage with the Real. If we want to make good on taking hold of the Anthropocene to limit the damage, or to at least go through this epoch together as a unified humanity, media is the necessary starting point.

NOTES

1. Christopher L. Salter, "The Cowboy and the City: Urban Affection for Wilderness," *Landscape* 27, no. 3 (1983): 43.
2. Salter, "Cowboy," 44.
3. Salter, "Cowboy," 45.
4. Salter, "Cowboy," 44.
5. Karl Marx, *Capital: A Critique of Political Economy, Volume One* (London: Penguin Books, 1990 [1867]), 272.
6. Louis Althusser, "Ideology and Ideological State Apparatuses (Notes Towards an Investigation)," in *Mapping Ideology*, edited by Slavoj Žižek (London: Verso, 2012 [1971]), 126.
7. Stuart Hall, "Signification, Representation and Ideology," *Critical Studies in Media Communication* 2, no. 2 (1985): 100, his emphasis.
8. Stuart Hall, "Encoding/Decoding," in *Media Studies: A Reader*, edited by Paul Marris and Sue Thornham, 51–61 (New York: New York University Press, 2000 [1973]).
9. Salter, "Cowboy," 45.
10. Karl Marx and Friedrich Engels, "Manifesto of the Communist Party," in *Karl Marx: The Political Writings* (London: Verso, 2019 [1850]), 92.

11. Slavoj Žižek, *The Sublime Object of Ideology* (London: Verso, 2008 [1989]), 142, his emphasis.

12. Jacques Lacan, *The Four Fundamental Concepts of Psychoanalysis*, translated by Alan Sheridan (New York: W.W. Norton and Company, 1998 [1978]), 168.

13. Henri Bergson, *Time and Free Will: An Essay on the Immediate Data of Consciousness* (Mineola: Dover, 2001 [1913]), 104.

Bibliography

Adorno, Theodor and Max Horkheimer. "The Culture Industry: Enlightenment as Mass Deception." In *Dialectic of Enlightenment*, 120–167. London: Verso, 1997 [1944].

Althusser, Louis. "Ideology and Ideological State Apparatuses (Notes Towards an Investigation)." In *Mapping Ideology*, edited by Slavoj Žižek, 100–140. London: Verso, 2012 [1971].

Apatow, Judd, dir. *The 40-Year-Old Virgin*. 2005; Universal City: Universal Pictures. Streaming.

Areddy, James T. "China Rules Out Animal Market and Lab as Coronavirus Origin." *The Wall Street Journal*. May 26, 2020, https://www.wsj.com/articles/china-rules-out-animal-market-and-lab-as-coronavirus-origin-11590517508.

Asayama, Shinichiro, Seita Emori, Masahiro Sugiyama, Fumiko Kasuga, Chiho Watanabe. "Are We Ignoring a Black Elephant in the Anthropocene? Climate Change and Global Pandemic as the Crisis in Health and Equality." *Sustainability Science* 16 (2021): 695–701.

Avildsen, John G., dir. *The Karate Kid*. 1984; Los Angeles: Columbia Pictures. Streaming.

Baudrillard, Jean. *The Gulf War Did Not Take Place*. Sydney: Power Publications, 2009 [1991].

Baudrillard, Jean. *Simulacra and Simulation*. Translated by Sheila Faria Glaser. Ann Arbor: The University of Michigan Press, 1994 [1981].

Bentham, Jeremy. "An Introduction to the Principles of Morals and Legislation." In *Utilitarianism and Other Essays*. Edited by Alan Ryan. London: Penguin Books, 2004.

Bergson, Henri. *Time and Free Will: An Essay of the Immediate Data of Consciousness*. Translated by F.L. Pogson. Mineola: Dover Publications, Inc., 2001 [1913].

Berkes, Fikret. *Sacred Ecology, Third Edition*. New York: Routledge, 2012.

Bernstein, Aaron. "Coronavirus, Climate Change, and the Environment: A Conversation on COVID-19 with Dr. Aaron Bernstein, Director of Harvard Chan

C-CHANGE." *Harvard T.H. Chan School of Public Health*, https://www.hsph.harvard.edu/c-change/subtopics/coronavirus-and-climate-change/.

Blackmon, David. "So, We're Back to 'Peak Oil' Again." *Forbes* Sept. 23, 2020, https://www.forbes.com/sites/davidblackmon/2020/09/23/so-were-back-to-peak-oil-again/?sh=9d8488e2c9a6.

Böhm, Steffen, Campbell Jones, Chris Land, and Mat Paterson, "Introduction: Impossibilities of Automobilities." *The Sociological Review* 54, no. s1 (2006): 8.

Borges, Jorge Luis. "On Exactitude in Science." In *Jorge Luis Borges: Collected Fictions*, New York: Penguin Books, 1998 [1946].

Boyle, Danny, dir. *28 Days Later*. 2002; Los Angeles: Fox Searchlight Pictures. Streaming.

Burton, Tim, dir. *Pee-Wee's Big Adventure*. 1985; Burbank: Warner Bros. Pictures. Streaming.

Butler, Judith. *Gender Trouble: Feminism and the Subversion of Identity*. New York: Routledge, 1990.

Chin, Peter. "The Problem with 'Mad Max' Feminism." *Christianity Today* June 3, 2015. http://www.christianitytoday.com/ct/2015/june-web-only/problem-with-mad-max-feminism.html.

Collet-Serra, Jaume, dir. *The Commuter*. 2018; Santa Monica: Lionsgate. Streaming.

Coté, Charlotte. *Spirits of our Whaling Ancestors: Revitalizing Makah and Nuu-chah-nulth Traditions*. Seattle: University of Washington Press, 2010.

Cresswell, Tim. *On The Move: Mobility in the Modern Western World*. New York: Routledge, 2006.

Crutzen, Paul J. and Eugene F. Stoermer. "The 'Anthropocene'." *Global Change Newsletter*, no. 41 (2000): 17–18.

Dant, Tim. "The Driver-car." *Theory, Culture, & Society* 21, no. 4–5 (2004): 61–79.

de Certeau, Michel. *The Practice of Everyday Life*. Berkeley: University of California Press, 1988.

de Coning, Alexis. "Recouping Masculinity: Men's Rights Activists' Responses to Mad Max: Fury Road." *Feminist Media Studies* 16, no. 1 (2016): 174–176, DOI: 10.1080/14680777.2016.1120491.

de Hartog, Jeroen Johan, Hanna Boogaard, Hans Nijland, and Gerard Hoek, "Do the Health Benefits of Cycling Outweigh the Risks?" *Environmental Health Perspectives*, 118, no. 8 (2010): 1109–1116.

del Toro, Guillermo, dir. *Pacific Rim*. 2013; Burbank, CA: Warner Bros. Pictures. Streaming.

Deleuze, Gilles. *Bergsonism*. Translated by Hugh Tomlinson and Barbara Habberjam. New York: Zone Books, 2006 [1966].

Deleuze, Gilles. *Cinema 1: The Movement Image*. Translated by Hugh Tomlinson and Barbara Habberjam. Minneapolis: University of Minnesota Press, 1986 [1983].

Deleuze, Gilles and Félix Guattari. *A Thousand Plateaus: Capitalism and Schizophrenia*. Translated by Brian Massumi. Minneapolis: University of Minnesota Press, 2016 [1980].

Deleuze, Gilles and Félix Guattari. *Anti-Oedipus: Capitalism and Schizophrenia.* Translated by Robert Hurley, Mark Seem, and Helen R. Lane. New York: Penguin Books, 1977 [1972].

Deleuze, Gilles and Claire Parnet. *Dialogues II, Revised Edition.* Translated by Hugh Tomlinson and Barbara Habberjam. New York: Columbia University Press, 2007 [1977].

Deffeyes, Kenneth S. *Hubbert's Peak: The Impending World Oil Shortage.* Princeton, NJ: Princeton University Press, 2002.

Dery, Mark. "'Always Crashing in the Same Car': A Head-on Collision with the Technosphere." *The Sociological Review* 54, no. s1 (2006): 223–239. DOI: 10.1111/j.1467-954X.2006.00645.x.

Dowling, Robyn and Catherine Simpson. "'Shift-the Way You Move': Reconstituting Automobility." *Continuum* 27, no. 3 (2013): 421–433. DOI: 10.1080/10304312.2013.772111.

Eisenstein, Sergei. *Film Form: Essays in Film Theory.* Translated by Jay Leyda. New York: Harcourt, Brace, and World, Inc., 1949.

Fisher, Mark. *Capitalist Realism: Is There No Alternative?* Winchester: Zero Books, 2009.

Foucault, Michel. *The History of Sexuality: Volume 1: An Introduction.* Translated by Robert Hurley. New York: Vintage, 1990 [1978].

Foucault, Michel. *Discipline and Punish: The Birth of the Prison.* Translated by Alan Sheridan. New York: Vintage Books, 1995 [1975].

Foucault, Michel. "Truth and Power." In *Power/Knowledge: Selected Interviews & Other Writings 1972-1977*, edited by Colin Gordon, 109–133. New York: Vintage Books, 1980 [1977].

Foucault, Michel. *The Archaeology of Knowledge and the Discourse on Language.* Translated by A.M. Sheridan-Smith. New York: Pantheon Books, 1972 [1969].

Furness, Zack. *One Less Car: Bicycling and the Politics of Automobility.* Philadelphia: Temple University Press, 2010.

Glaeser, Edward. *Triumph of the City: How Our Greatest Invention Makes Us Richer, Smarter, Greener, Healthier, and Happier.* New York: Penguin Books, 2011.

Gramsci, Antonio. *Selections from the Prison Notebooks.* Edited and translated by Quintin Hoare and Geoffrey Nowell Smith. New York: International Publishers, 1971.

Groth, Paul. "Frameworks for Cultural Landscape Study." In *Understanding Ordinary Landscapes*, edited by Paul Groth and Todd W. Bressi. New Haven: Yale University Press, 1997.

Grove, Richard. *Green Imperialism: Colonial Expansion, Tropical Island Edens, and the Origins of Environmentalism, 1600-1860.* Cambridge: Cambridge University Press, 1995.

Guerlac, Suzanne. *Thinking in Time: An Introduction to Henri Bergson.* Ithaca: Cornell University Press, 2017.

Halberstam, Jack. *Feminine Masculinity.* Durham: Duke University Press, 1998.

Halberstam, Jack. *Skin Shows: Gothic Horror and the Technology of Monsters.* Durham: Duke University Press, 1995.

Hall, Stuart. "Signification, Representation, Ideology: Althusser and the Post-Structuralist Debates." *Critical Studies in Mass Communication* 2, no. 2 (June 1985): 91–114.

Hall, Stuart. "Encoding/Decoding." In *Media Studies: A Reader*, edited by Paul Marris and Sue Thornham, 51–61. New York: New York University Press, 2000 [1973].

Haraway, Donna J. *Staying with the Trouble: Making Kin in the Chthulucene*. Durham: Duke University Press, 2016.

Harvey, David. *A Companion to Marx's Capital: The Complete Edition*. London: Verso, 2018 [2010].

Harvey, David. *The Condition of Postmodernity*. Malden: Blackwell Publishing, 1990.

Hawk, Steve. "The Fury Behind Fury Road." *Sierra* April 9, 2015. http://sierraclub.org/sierra/2015-3-may-june/feature/fury-road-all-your-darkest-environmental-nightmares-come-true.

Hersher, Rebecca. "Earth is Barreling Toward 1.5 Degrees Celsius of Warming, Scientists Warn." National Public Radio. May 26, 2021. https://www.npr.org/2021/05/26/1000465487/earth-is-barreling-toward-1-5-degrees-celsius-of-warming-scientists-warn.

Horton, Dave. "Environmentalism and the Bicycle." *Environmental Politics* 15, no. 1 (2006): 41–58.

Huang, Erin Y. *Urban Horror: Neoliberal Post-Socialism and the Limits of Visibility*. Durham: Duke University Press.

Intergovernmental Panel on Climate Change. "Summary for Policy Makers." In *Global Warming of 1.5°C. An IPCC Special Report on the impacts of global warming of 1.5°C above pre-industrial levels and related global greenhouse gas emission pathways, in the context of strengthening the global response to the threat of climate change, sustainable development, and efforts to eradicate poverty*. Edited by V. Masson-Delmotte, P. Zhai, H.-O. Pörtner, D. Roberts, J. Skea, P.R. Shukla, A. Pirani, W. Moufouma-Okia, C. Péan, R. Pidcock, S. Connors, J.B.R. Matthews, Y. Chen, X. Zhou, M.I. Gomis, E. Lonnoy, T. Maycock, M. Tignor, and T. Waterfield, https://www.ipcc.ch/site/assets/uploads/sites/2/2019/05/SR15_SPM_version_report_HR.pdf.

Jackson, Mick, dir. *L.A. Story*. 1991; Culver City: TriStar Pictures. Streaming.

Joon-ho, Bong, dir. *Snowpiercer*. 2013; Seoul: CJ Entertainment. Streaming.

Klangwisan, Yael. "Mad Max Fury Road: Escaping the Phallic Economy of the Exodus." *The Bible and Critical Theory* 12 (2016): 75–86.

Lane, Anthony. "High Gear: 'Mad Max: Fury Road'." *The New Yorker* May 25, 2015. http://www.newyorker.com/magazine/2015/05/25/high-gear-current-cinema-anthony-lane.

Lacan, Jacques. *The Four Fundamental Concepts of Psychoanalysis*. Translated by Alan Sheridan. New York: W.W. Norton and Company, 1998 [1978].

Latour, Bruno. *Facing Gaia: Eight Lectures on the New Climatic Regime*. Cambridge: Polity Press, 2017.

Lefebvre, Henri. *The Production of Space*. Translated by Donald Nicholson-Smith. Malden: Blackwell Publishing, 1991 [1974].

Malm, Andreas. *Fossil Capital: The Rise of Steam Power and the Roots of Global Warming*. London: Verso, 2016.
Marx, Karl. *Capital: A Critique of Political Economy, Volume One*. London: Penguin Books, 1990 [1867].
Marx, Karl and Friedrich Engels. "The German Ideology: Part I." In *The Marx-Engels Reader, Second Edition*, edited by Robert C. Tucker, 146–200. New York: W.W. Norton and Company, 1978 [1846].
Marx, Karl and Friedrich Engels. "Manifesto of the Communist Party." In *Karl Marx: The Political Writings,* 61–92. London: Verso, 2019 [1850].
Massey, Doreen. *For Space*. London: Sage Publications, 2005.
McNeill, John Robert and Peter Engelke. *The Great Acceleration: An Environmental History of the Anthropocene since 1945*. Cambridge: Harvard University Press, 2014.
Miller, George, dir. *Mad Max: Fury Road*. 2015; Burbank, CA: Warner Bros. Pictures. Streaming.
Moore, Jason W. "The Capitalocene, Part I: On the Nature and Origins of Our Ecological Crisis" *The Journal of Peasant Studies* 44, no. 3 (2017): 594–630.
Moore, Jason W. *Capitalism in the Web of Life: Ecology and the Accumulation of Capital*. London: Verso, 2015.
Offenhartz, Jake. "Sometimes the City has to Destroy a Bike Lane Before it Can Reinstall a Bike Lane in the Same Place." *Gothamist*, November 4, 2019, https://gothamist.com/news/sometimes-city-has-destroy-bike-lane-it-can-reinstall-bike-lane-same-place.
Owens, David. *The Conundrum: How Scientific Innovation, Increased Efficiency, and Good Intentions Can Make Our Energy and Climate Problems Worse*. New York: Riverhead Books, 2011.
Pesses, Michael W. *Crawl Space: Driving over the Anthropocene in a Jeep*. PhD diss. Claremont Graduate University, 2020.
Pesses, Michael W. "Automobility, Vélomobility, American Mobility: An Exploration of the Bicycle Tour." *Mobilities* 5, no. 1 (2010): 1–24.
Philpott, William. *Vacationland: Tourism and Environment in the Colorado High Country*. Seattle: University of Washington Press, 2013.
Purdy, Jedediah. 2015. *After Nature: A Politics for the Anthropocene*. Cambridge: Harvard University Press, 2015.
Rojas-Rueda, David, Audrey de Nazelle, Marko Tainio, Mark J. Nieuwenhuijsen. "The Health Risks and Benefits of Cycling in Urban Environments Compared with Car Use: Health Impact Assessment Study." *BMJ* 343, no. 7819 (2011): 1–8.
Rosenthal, David M., dir. *How It Ends*. 2018; Los Gatos, CA: Netflix. Streaming.
Ruddiman, William F. "The Anthropogenic Greenhous Era Began Thousands of Years Ago." *Climatic Change* no. 61 (2003): 261–293.
Russo, Tom. "'Pacific Rim' is the heartfelt project from del Toro." *Boston Globe*. July 6 2013, https://www.bostonglobe.com/arts/movies/2013/07/06/pacific-rim-latest-heartfelt-project-from-guillermo-del-toro/aTU5X0wLpNaUe1GpbHqDUO/story.html.
Sadik-Khan, Janette. "The Bike Wars Are Over, and the Bikes Won." *New York*, March 8, 2016, https://nymag.com/intelligencer/2016/03/bike-wars-are-over-and-the-bikes-won.html.

Salter, Christopher L. "The Cowboy and the City: Urban Affection for Wilderness." *Landscape* 27, no. 3 (1983): 43–47.

Scanlon, Dan, dir. *Onward*. 2020; Burbank: Walt Disney Pictures. Streaming.

Sheller, Mimi. "Automotive Emotions: Feeling the Car." *Theory, Culture & Society* 21, no. 4/5 (2004): 221–242.

Shukin, Nicole. "The Mimetics of Mobile Capital." *The Sociological Review* 54, no. s1 (2006): 150–174. DOI: 10.1111/j.1467-954X.2006.00642.x.

Sieler, Cotten. *Republic of Drivers: A Cultural History of Automobility in America*. Chicago: The University of Chicago Press, 2008.

Smith, Kyle. "Why 'Mad Max: Fury Road' is the Feminist Picture of the Year." *The New York Post* May 14, 2015. http://nypost.com/2015/05/14/why-mad-max-fury-road-is-the-feminist-picture-of-the-year/.

Smith, Neil. *Uneven Development: Nature, Capital, and the Production of Space, Third Edition*. Athens: University of Georgia Press, 2008.

Speck, Jeff. *Walkable City: How Downtown Can Save America One Step at a Time*. New York: Farrar, Straus, and Giroux, 2012.

Sputore, Grant, dir. *I Am Mother*. 2019; Los Gatos, CA: Netflix. Streaming.

Stoler, Ann Laura. *Duress: Imperial Durabilities in Our Times*. Durham: Duke University Press, 2016.

United Nations. *Paris Agreement*. 2015, https://unfccc.int/files/essential_background/convention/application/pdf/english_paris_agreement.pdf.

Urry, John. "Inhabiting the Car." *The Sociological Review* 54, no. s1 (2006): 17–31. DOI: 10.1111/j.1467-954X.2006.00635.x.

Venter, Zander S., Kristin Aunan, Sourangsu Chowdhury, and Jos Leleiveld. "COVID-19 Lockdowns Cause Global Air Pollution Declines." *Proceedings of the National Academy of Sciences* 117, no. 32 (August 11, 2020): 18984–18990.

Virilio, Paul. *Speed and Politics*. Translated by Mark Pilizzotti. South Pasadena: Semiotext(e), 2006 [1977].

Wachowski Siblings, The, dirs. *The Matrix*. 1999: Burbank, CA: Warner Bros. Pictures. Streaming.

Wainwright, Joel, and Geoff Mann. *Climate Leviathan: A Political Theory of Our Planetary Future*. London: Verso, 2018.

Wark, McKenzie. *Molecular Red: Theory for the Anthropocene*. London: Verso, 2016.

Weart, Spencer R. *The Discovery of Global Warming, Revised and Expanded Edition*. Cambridge: Harvard University Press, 2008.

World Meteorological Organization. "New climate predictions increase likelihood of temporarily reaching 1.5°C in next 5 years." May 27, 2021, https://public.wmo.int/en/media/press-release/new-climate-predictions-increase-likelihood-of-temporarily-reaching-15-°c-next-5.

Žižek, Slavoj. *Violence*. New York: Picador, 2008.

Žižek, Slavoj. *The Plague of Fantasies*. London: Verso, 2008 [1997].

Žižek, Slavoj. *The Sublime Object of Ideology*. London: Verso, 2008 [1989].

Žižek, Slavoj. *Welcome to the Desert of the Real*. London: Verso, 2002.

Index

About the Author

Michael W. Pesses is a professor of geography at Antelope Valley College in Lancaster, California. His research interests primarily revolve around the concept of automobility, whether through its historical, social, ideological, or environmental aspects. Such research usually takes the form of consuming a variety of media or driving over things.

www.ingramcontent.com/pod-product-compliance
Ingram Content Group UK Ltd.
Pitfield, Milton Keynes, MK11 3LW, UK
UKHW020842290926
15470UKWH00004B/1506

9 781498 598194